Living with Technology

Living with Technology: Issues at Mid-Career

Lotte Bailyn

in collaboration with
Edgar H. Schein

The MIT Press
Cambridge, Massachusetts, and London, England

This book was set in Fototronic Zenith by The Colonial Cooperative Press Inc. and printed and bound by The Murray Printing Company in the United States of America.

Library of Congress Cataloging in Publication Data

Bailyn, Lotte.
Living with technology.

Bibliography: p. 123
Includes index.
1. Engineers. 2. Engineering—Social aspects. 3. Massachusetts Institute of Technology—Alumni. I. Schein, Edgar H. II. Title.
TA157.B29 306′.3 80-23970
ISBN 0-262-02153-6

To the memory of
Paul F. Lazarsfeld

Contents

	Preface	**xi**
1	**Introduction: Midcareer Issues in Technology-Based Careers**	**1**
	Pluralism: A Recurrent Theme	4
	The Research Sample	6
	A Guide to the Book	7
	Notes	9
2	**Occupational Roles**	**10**
	Eleven Occupational Categories	15
	Notes	18
3	**Career Patterns and Values**	**19**
	The Two Major Career Patterns	19
	Differences in Job Values	22
	Socialization or Preselection?	28
	Summary	30
	Notes	31
4	**Reactions to Work**	**33**
	Work Involvement	34
	Organizational Position	36
	Four Types of Technology-Based Careers	38
	Summary	43
	Notes	44
5	**Individual Orientations and Organizational Career Paths**	**47**
	Individual Orientations to Work	48
	The Fit Between Individual Orientations and Career Type	50

Four Career Types: A Summary Profile 53
Notes 55

6 **The Women Graduates** **57**
Career Paths 58
Midlife Career Styles 61
Conclusion: Future Trends? 64
Notes 66

7 **Implications and Conclusions: Toward New Organizational Roles** **67**
Summary of Findings 67
Implications for the Organization 70
Concluding Note 78
Notes 79

Appendixes

A **Supplementary Tables** **83**

B **Population and Sample by Year of Graduation** **89**

C **Occupational Success** **93**

D **Work Involvement** **97**

E **Orientations to Work** **107**

F **The Questionnaire** **111**

Bibliography **123**

Index **129**

Tables

2.1 Basic Occupations of the Alumni Sample 13

2.2 Eleven Occupational Categories 16

3.1 Career Patterns Based on Early-Career Events 20

3.2 List of Job Values, with Percentage Rating Each "Very Important" 24

3.3 Percentage of "Successful" Respondents Rating Given Job Characteristics "Very Important" 26

3.4 Percentage of "Successful" Alumni in Congruent and Incongruent Occupations Rating Given Job Characteristics "Very Important" 30

4.1 Work Involvement by Occupational Role 36

4.2 Success Aspirations of Occupants of High and Low Organizational Positions 37

4.3 Four Career Types 38

4.4 Occupational Value Norms and Characteristic Career Stance of Four Career Types (Including Data on Work Involvement) 40

4.5 Reactions to Work by Occupants of Four Career Types 43

5.1 Distribution of Unambiguously Oriented Individuals 49

5.2 Orientations in Four Career Types 49

5.3 Organizational Evaluation: Percentage with Above-Average Incomes in Each Career Type and Orientation 50

5.4 Individual Satisfaction: Percentage "Very Satisfied" with Their Jobs in Each Career Type and Orientation 51

5.5 Percentage "Very Satisfied" with Their Jobs in Career Types E/H and E/L According to Relative Income and Orientation 53

6.1 Career Style and Family Status Among Alumnae 59

6.2 Career Style and Reactions to Work of Employed Alumnae 62

7.1 (Chart) Midcareer Organizational Roles 72

A.1 Occupational Roles by Year of Graduation 83

A.2 Some Background and Current Characteristics of Respondents in Eleven Occupational Roles 84

A.3 Percentage of "Successful" Respondents Rating Other Job Characteristics "Very Important" 86

B.1 Representation of Undergraduate Majors in Population and Sample, by Class 90

C.1 Occupational Success as Measured by Combination of Responses to the Two Questions of Item 19 94

C.2 Perceived Success in Career Patterns and Occupational Roles 95

D.1 Final Variables Used in Index of Work Involvement 97

D.2 Distribution on Final Measure of Work Involvement, Based on Average of Three 5-Point Scales 100

D.3 Correlations of Work-Involvement Index with Components 101

D.4 Some Differentiating Characteristics Between Groups with Extremely High and Extremely Low Work-Involvement Scores 102

D.5 Work Involvement in Relation to Intrinsic and Extrinsic Job Characteristics 105

E.1 Technical Orientation Determined from Local/ Cosmopolitan Distinction and Agreement with "Change of Work" Statement 108

Preface

The study from which this book developed was an attempt to understand the career patterns and reactions to work of MIT graduates who were well established in their careers. It was undertaken out of both a specific interest in the fate of MIT alumni and a more general interest in the development and patterning of careers based on technical education. Such careers, we found, are experienced in widely different ways. Some are highly satisfactory, others surprisingly alienating, and the differences seem to lie as much in the positions and roles of technical jobs within organizations as in the people who pursue them. These differences in career satisfaction, their relation to effectiveness, and their meaning for the individuals and for the organizations in which they work are the main subjects of this book.

In studying the careers of MIT alumni, we had no illusion that they represent the general occupational distribution in this country, or even its most educated part. We were not seeking a broadly representative sample. Attendance at MIT defines a carefully selected group of people whose distinctiveness is reinforced by the education they receive there. By the time they leave MIT they are highly trained and skilled professionals, deeply involved, psychologically and intellectually, in science and technology. Such preparation leads to major technical and managerial roles in both public and private organizations; understanding the forms that these careers take and the reactions of the people who pursue them is important to an understanding of our society.

Concentration on this MIT group has a further advantage. Studying alumni of a single institution eliminates differences in undergraduate education and all the selective factors associated with that experience. Career differences observed in such a relatively homogeneous group are less likely to result from individual differences present as people enter employment, and more likely to reflect the actual conditions of technical work. Implications for organizational policies drawn from such findings should therefore provide a more accurate guide to employing organizations than suggestions based on differences in a group where initial professional training is more varied.

The main theme that emerges from the findings of this study is that the varied orientations to work at mid-career require a *variety* of roles for technically trained employees. Managing this pluralism is a major and increasing challenge for organizations. Too often the need for multiple career paths is not even recognized by employing organizations, which perpetuate policies based on simplistic and monolithic assumptions about technical careers and organizational requirements. That such assumptions are outmoded, and that they result in costly dysfunctions for organization and employee alike, are documented in this book. The implications of multiple career patterns for the management of technical personnel are considered directly and in detail in the final chapter. Readers will find there a full discussion of creative responses that organizations might make once the nature of the issues involved is understood. Anyone particularly concerned with management problems may wish to turn to that chapter first, before reading the documentation and analysis that led to the recommendations presented there.

A word should be said about the use of empirical data. The analytic strategy used in this book follows what I have called the cognitive process of research (Bailyn 1977c). It is not a procedure to test preformulated hypotheses; rather, it is an attempt to gain insight by identifying and developing new concepts as they emerge from data. In this study, such concepts emerge from the progressive refinement of categories of people and of the characteristics of the jobs they hold, and from detailed comparisons of the responses of different groups to different kinds of work. This approach can tell us a great deal about the relations between people and their work. It can pinpoint the areas of difficulty, and it can help isolate the positions that, with respect to particular types of employees, are most problematic and most in need of rethinking. This kind of analysis—which allows one to draw conclusions that are solidly grounded in reality—is not often amenable to statistical reduction, and thus is necessarily painstaking and detailed. But if readers are to be able to evaluate the interpretations drawn from the data, these details must be presented in full.

The study was begun with Edgar H. Schein under a grant from the Carnegie Commission on the Future of

Higher Education. It was supported also by the Office of the Undergraduate Professor at MIT, a position Professor Schein held at the time. The help of the Carnegie Commission and the Undergraduate Professorship is gratefully acknowledged. My work on the project, which began before the basic questionnaire was drafted, was supported by research funds from the Sloan School of Management at MIT. I am grateful to Dean Abraham Siegel for making this help available to me.

Professor Schein contributed particularly the original occupational classification and the analysis of value syndromes. His subsequent involvement in the project, acknowledged on the title page, was part of a continuing collaboration at the Sloan School, where problems such as those discussed in the pages that follow are being tackled in a variety of ways by the Organization Studies Group.

Many other people have participated in this research, including a generation of students who, as members of my seminar on research methods, have helped in the analysis of the data. Of these, Marc Gerstein should be mentioned specifically. Two people contributed greatly to the administration of the project: Dany Siler handled this task from the study's inception through its first results; both she and her successor, Ruth Guetta, were instrumental in pushing the work forward, and their care and good cheer are gratefully acknowledged. I am grateful, also, to Brenda Venuti, who managed the production of the manuscript with efficiency and care, and to Cheryl Fennessey, who typed and retyped without complaint.

John Van Maanen and Bernard Bailyn both read the entire manuscript, and their thoughtful and extensive comments, based on different points of view, were most helpful. And Paul F. Lazarsfeld, to whose memory this book is dedicated, contributed in a number of ways—some obvious, some not so obvious.

Finally, I am indebted to those alumni who responded to the questionnaire, who took the time and care necessary to provide the information on which this study is based. I hope that if they read this book they will feel that their efforts were not in vain.

Lotte Bailyn
Cambridge, Massachusetts

Living with Technology

1 Introduction: Midcareer Issues in Technology-Based Careers

The plight of the alienated worker has concerned social analysts for more than a century. Fragmented work, monotonous and meaningless jobs, lack of autonomy, and lack of control over working conditions are among the factors contributing to this condition. The situation of the professional worker or manager who enters the occupational world with the benefit of at least four years of advanced education seems strikingly different. It has been assumed and argued that occupations available to this highly educated group are challenging and fulfilling throughout life, and that in these careers one can be the master rather than the slave of one's work. But is this true? Evidence is mounting that negates this assumption. There are executives who have midcareer crises, there are engineers who become bored with their work, and there are people who leave high positions in organizations to start alternative careers. Events such as these point to an underlying problem as yet largely unexplored, indeed not yet fully identified.

To be sure, partial explanations for these career problems lie readily at hand. Such events can be viewed as responses to a changing social order in which technology plays an ever more complicated role. They may also be partially explained by the patterns of personal development and by the changes in people's expectations about the kinds of lives they want to lead. However, such societal and developmental explanations pertain to all technically trained people in a given era. To find out why some flourish and others become alienated, one must look at the way specific technical careers evolve as a function of both individual orientations and organizational procedures. The present study is an effort to provide such an analysis. It examines the careers of a group of people trained at one elite technical institute some 10–20 years after their graduation.

The development of a career depends on the fit between an individual's needs, talents, and values and the demands of successive jobs. It is a complex process because these personal attributes are themselves shaped by work experiences. In looking at a career at any particular point, one views only one stage in the process of interactive accommodation between the individual and the work environment. But detailed consideration, at such a stage, of a person's level of work involvement, extent of job satisfaction, and perception of occupational success or failure yields some insight into the underlying developmental process.

The study reported here deals with a group of technically trained people who have reached mid-career, a stage of life in which there is currently much interest (see Rapoport 1970; Sofer 1970; Pearse and Pelzer 1975). Moreover, it is a group hard hit by changing values. The respondents are approaching mid-career at a time when changes in society are exacerbating the effects on them of this period in their lives. In the 1950s, when the respondents were at MIT, science and technology were hailed as offering solutions to most of the world's problems. They were considered central to national unity and social betterment, and their study was thought to lead to almost certain personal success. But today, the value of technical progress—the basic underpinning of most of the careers of these alumni—is being questioned. "It would not have been possible to convince me that this would happen twenty years ago," writes a technical manager who graduated in 1951; "however, I am now inclined to believe that technology has created more problems than it has solved, and is—in that sense—at least an insufficient answer and possibly a dead end." And as society begins to impose social and legal constraints on new knowledge, the unique value of the scientific method itself is increasingly under question.[1] In the words of one 1955 graduate, "Our society is sterilizing a generation of scientific creativity and causing bitter disillusionment in literally thousands of engineers and scientists. . . ."

These alumni are also confronted with challenges to the accepted meaning of achievement and success. The usual rebellion of youth against parental values is being turned into a more enduring challenge (Yankelovich 1972). It has meshed with the egalitarian movement for minorities and women, which has gone beyond matters such as voting rights, equal pay, or even equal access to opportunities and now challenges some of the most basic assumptions underlying occupational roles in our society.

Thus, we study an especially significant group at a time

when they may feel the very basis of their lives coming into question. These people (particularly the men) decided relatively early in their lives to pursue some of the most demanding occupations available. At MIT they pursued a rigorous course of technical study and entered the world of work expecting far more from it than the mere satisfaction of instrumental needs. They probably expected to find jobs at the highest rungs of the occupational ladder, jobs in which a close tie between work and one's basic sense of self would be likely to exist (Wilensky 1964). The assumption was that career success and achievement would ensure a satisfactory life. Family and community would naturally be of great importance, but neither would be expected to detract from work as the central focus of a person's life.

For some, of course, these expectations were met. But not for all. Many have discovered (some to their surprise) that these early aspirations, even when met, are not as satisfactory as had been anticipated. One alumnus reports that being "successful" and "being paid a lot" are "not all that impressive." He has tried to quit his job "and find an occupation which would have more relevance and influence on social problems," and comments movingly on what educational institutions like MIT might do to help alumni change from the occupations for which they were originally trained to other, more satisfying ones. Another respondent, who has "fairly recently attained independent means," proposes "to let my involvement with science and technology run out" and to let his "principal creative energies" go into "sculpture, lapidary work, etc. which gives me much more personal satisfaction than science or engineering has ever done." And a 40-year-old director of marketing, married for the first time at the age of 35, writes

> Now that I find myself with a family, much to my amazement I find that I really enjoy being with my family and almost resent any intrusion on my home life. I would rather spend my time with my family than on committees; would rather be home than on the road; and I find that my newly acquired domesticity has tamed my desire to be a leader of anything other than my family.

These individual reactions give a flavor of the recent lives of some of the respondents. But what are the overall patterns of their careers? Are there systematic ways of understanding their reactions which could be used to devise more satisfactory career paths for technical personnel? These are the questions addressed in this book.[2] The answers are not universal, but neither are they individually unique. Rather, they depend on looking at both people and jobs in a patterned but differentiated way.

Pluralism: A Recurrent Theme

Recent writing on adult life (for example, Sheehy 1976; Vaillant 1977; Gould 1978; Levinson et al. 1978) has attuned us to associate certain ages with particular sets of problems. The "terrible twos" and "equable fives" popularized in the writings on childhood have been extended to the "trying twenties," "catch-30," and the "mid-life transition." But though we only recently have given names to these so-called stages of adult development, the paths of most technical careers in organizations are based on a similar supposition about the existence of predictable and universal transitions in adult lives. "In industry," writes a 1951 alumnus, "an engineer who stays an engineer past the age of 35 or 40 is a failure." Presumably, all engineers, or at least the good ones, should shift at mid-career from technical to human concerns. They should enter a stage of "generativity" (Erikson 1959) and should be content to leave the technical work to younger and more "up-to-date" employees.

But career paths based on such monolithic presumptions do not necessarily fit the orientations and concerns of the people who pursue them. Indeed, there is much evidence that they do not:

> Supposedly my company has a dual ladder system for advancement, the technical and the managerial. I have been most attracted to, and have concentrated my efforts towards, the technical path, but now, technical contributions comparable to my own can be bought at a lower price from a recent engineering graduate. Presumably, my years of technical experience with the company have little or no economic value to them.
>
> —senior development engineer, 42 years old

> What a shock it is to an engineer to discover that the man who has developed leadership talents and has the ability to communicate is more promotable than the one who is primarily technically oriented. Even those entering the research lab are faced with this fact.
>
> —assistant vice president, 42 years old

Why then does the myth that all engineers should eventually become managers persist? Partly it is because the people who do want to move into positions of power, and who are successful in doing so, are the ones who define the rewards in their organizations. With no explicit strictures to the contrary, they proceed on the assumption that all high performers have interests and capacities similar to theirs. To the extent that this assumption is false they thus create unnecessary dissatisfaction among employees and reduce the effectiveness of the technical work force.

Many technical educators, on the other hand, subscribe to the exactly opposite monolithic assumption, that technically trained graduates will be able and willing to pursue lifelong technical careers. They then respond to problems encountered in those careers by producing courses and other opportunities for technical updating. Again, the assumption probably fits those who hold it; engineering and science professors work in an environment in which they are more able than their industrial counterparts to pursue satisfactory careers centered on technical work. But it does not necessarily fit the organizational reality into which students not headed for academic careers graduate:

My MIT undergraduate education provided me with great ability to solve problems of a strictly technical nature. However, it was weak in providing me with the training necessary to handle many typical industrial situations.
—project manager, class of 1951

I feel that MIT prepared me to be able to do the technical things necessary to be a technical leader of considerable technical capability. However, I feel that I have been completely unprepared for the job of recognizing and doing the nontechnical things (social and political) necessary to enter the leadership circle within the organization of a very large company.
—engineer/group leader, class of 1955

To introduce management-related, or people-oriented, or "real-world" experiences into technical education, as is increasingly being done, is only part of the answer. It is even more critical that managers of technically trained personnel learn to eschew monolithic assumptions and to accept the wide variation in interests and capacities of technical graduates at mid-career. They would then be able to provide the highly differentiated organizational responses that such variety calls for. The data to be presented support the need for more pluralistic assumptions. They show that wide variations in reactions to work exist even in a relatively homogeneous and selected group.

The Research Sample

The research population consists of the MIT classes of 1951, 1955, and 1959 (N=2,223). Almost two-thirds of the graduates in these classes studied in the School of Engineering, particularly in the departments of electrical engineering, mechanical engineering, and chemical engineering. About one-sixth studied in the School of Science, almost half of whom graduated in physics with the rest pretty evenly divided among the other science disciplines. Of the remainder, most (about one-eighth of the total) graduated from the School of Management, with the balance from the School of Architecture and the School of Humanities and Social Sciences. There were 22 female graduates in these classes.

Every alumnus in these classes was sent in 1970 a questionnaire (reproduced as appendix F), which asked for a detailed objective job history, opinions about educational experiences while at MIT, current work attitudes and values, and self-perceptions along a number of dimensions. With the aid of one followup postcard we obtained 1,366 usable responses, a response rate of 61%. (Of the 22 women, 15 responded.) As may be seen from appendix B, the distribution of the undergraduate majors in this 61% is not significantly different from that of the total population. To the extent that it is possible to check, therefore, the sample is representative of the population.

As a group, the respondents were gifted in quantitative skills from a relatively early age and came to MIT for professional training in science and technology. They were destined for the key technology-based jobs in our society, for the positions on which our national welfare must depend. As the MIT catalogs of the time put it, the Institute "aims to give its students such a combination of humanistic, scientific, and professional training as will fit them to take leading positions in a world in which science, engineering, and architecture are of basic importance," to prepare them "to become practicing engineers or architects, investigators, business executives, or teachers" (MIT

Catalog, 1951). The purpose of the study is to investigate these careers in detail and thereby to illuminate the midcareer issues, both individual and organizational, of technology-based career paths.

A Guide to the Book

Chapters 2–5 deal with the careers of the 1,351 men in the sample. Chapter 2 details their midcareer occupational roles. Although the variety of positions is wide, most (over 85%) fall into eleven technology-related professorial, managerial, or staff roles in government, business, universities, or laboratories. The chapter discusses some of the differences among these eleven groups of alumni in socioeconomic background, in employment sector and income, and in current family characteristics. It provides the basic occupational categorization for the analysis. Readers not concerned with the specific midcareer occupations of the respondents may wish to skip over this chapter.

Chapter 3 describes the two major career patterns that emerge from an examination of the respondents' educational and early-career experiences: an engineering-based pattern, which includes the technical and business roles of trained engineers, and a scientific-professional pattern, which comprises professorial, managerial, and staff-scientist roles that are anchored in the disciplines of basic science. An important point that emerges from the analysis in this chapter is that these two basic career patterns can be differentiated by the characteristic value norms associated with them. The scientific-professional pattern is associated with a professional value syndrome, whereas an organizational value syndrome is more characteristic of the engineering-based career pattern. The final analytic section of the chapter shows that these value norms were formed in large part by the occupational experiences of the alumni, and were not necessarily present when occupational choices were made.

In chapter 4 the focus shifts from occupational role to individual reactions to work. This chapter deals primarily with the respondents' work involvement, but discusses also their perception of their own career success and their satisfaction with their jobs. The initial comparison of work involvement among the occupational roles of each career pattern shows that for this more detailed analysis a further differentiation is necessary. Specifically, within each career pattern certain occupational roles are associated with high positions in organizations and others with low positions, and this distinction is tied to the individual's level of success aspiration. It turns out that work involvement is as much associated with organizational position as it is with whether the career pattern is engineering-based or scientific-professional. The chapter, then, concludes with a discussion of four basic career types: engineering-based careers in high and low organizational positions and scientific-professional careers in high and low organizational positions. Each of these has associated with it a particular combination of value norms and success aspirations and a distinct pattern of modal reactions.

These findings lead directly to the analysis in chapter 5, which asks whether there are certain types of people who do not have these modal reactions. To answer this question it is necessary to classify the respondents by their individual orientations, initially by whether or not their primary involvement is with their work or with other aspects of their lives, and then, if their orientation is toward work, by whether it is more toward the technical or the human aspects of work. It is then possible to examine the fit between a given individual orientation and the person's career type. The chapter looks at this question from both the organization's point of view (Which orientation is most rewarded?) and the individual's (Which provides the greatest satisfaction?). It ends with a summary profile of the four career types, which should be of interest even to readers not concerned with the details of the analysis.

Chapter 6 considers the women who graduated in the same MIT classes. Because of the small number involved, it necessarily deals with the data in a very different way. This group is probably even more unusual than the men, as only 1% of each class at that time was female. (In the late 1970s the figure was close to 20%.) But the lives of these highly selected women are of particular interest because they suggest career accommodations that seem to be becoming more prevalent among recent college graduates of both sexes.

Finally, in chapter 7, the strands are all pulled together. The chapter starts with a short summary of findings and then deals in detail with their implications. In particular, it considers the question of what can be learned from the careers of these alumni about how technical careers in organizations should be managed, and suggests new organizational roles and changes in personnel policies that would heighten the effectiveness of these employees and increase their satisfaction with their work.

Though the chapters contain substantive details, the most technical material is presented in the appendixes. Since it is not always easy to base inferences about career questions on retrospective questionnaires, these appendixes further an incidental purpose of this book: to illustrate how data from a cross-sectional survey can be used to explore such complex questions.

Notes to Chapter 1

1. See the Spring 1978 issue of *Daedalus* on the limits of scientific inquiry.

2. Parts of this picture are already available. Some basic occupational data on the group have been published previously (Bailyn and Schein 1972), selected problems of specific subgroups of engineers and general managers have been reported (Plovnick 1972; Schein 1972; Madrazo 1974; Johnson 1975; Bailyn 1977a), and some analyses have been made of the interrelation of work and family in the lives of this alumni group (Bailyn 1973, 1977a).

2 Occupational Roles

The respondents in this study share a number of general characteristics. First, all are graduates of a technological institute in which a core of mathematics and science courses was an absolute undergraduate requirement, even among those who majored in "nontechnical" fields. And though some alumni now have jobs that seem unrelated to their undergraduate majors (such as a physics graduate doing photography), most of the careers under study stem directly from the respondents' initial technical training.[1]

Second, the decision to come to MIT presumably reflects a particular pattern of talents, motives, and values. The desire to be educated in science and technology implies an early commitment to a particular range of fields. These fields attract people who have certain personality traits and needs that distinguish them from those in the humanities. Not surprisingly, those destined for science and engineering show an early inclination more toward scientific, mechanical, and quantitative activities than toward aesthetic pursuits (Sternberg 1955; Hudson 1967). Compared with those in humanities they would rather deal with objects than with human beings (Roe 1957; Rosenberg 1957; Perrucci and Gerstl 1969). Scientists and engineers have a high need for achievement (Izard 1960; Dipboye and Anderson 1961). They are concerned with order and stability (Moore and Levy 1951; Steiner 1953; Izard 1960; Roe 1961; Perrucci and Gerstl 1969) and are less flexible than those in humanities. They are also more convergent than divergent in cognitive style (Hudson 1967; Kolb and Goldman 1973). There is no direct evidence of the personalities, cognitive styles, and values of the respondents when they entered MIT, but their predilections, reinforced by the undergraduate curriculum, most likely fit this pattern. Though subsequent to graduation their careers followed a variety of paths, this common base contributes to the occupational positions they hold at mid-career.

In order to develop an occupational classification for such technology-based jobs, we used the title of each respondent's current job, the functions he performs in that job, and the type of organization in which he works.

These objective facts were augmented where necessary by information from the job history.

A basic decision had to be made about the meaning of "management." We did not consider anyone a manager whose job did not include responsibility for and supervision of people. A portfolio manager in an investment firm, for example, would not have met this criterion, even though the job title includes the word "manager." He would have been included, rather, in a "staff" category.[2] In general, the use of "management" in this study is based on an evaluation of the extent to which a particular job involves responsibility for and supervision of people. It does not reflect the use of the term in a job title, nor does it depend on any assessment of whether a particular job represents a "line" position.

Within this category of management, we wanted to differentiate unambiguously among those respondents whose roles were clearly entrepreneurial; those who were in general management; and those who had a particular function, technical or nontechnical. We did not want to confuse any of these categories with that of the manager of a family concern, where it would be difficult to judge how much of his performance was the result of his own achievement and how much reflected the initial family position. Nor did we want to confuse first-level supervision with management. In order to meet these criteria, the managerial categories were defined in the following way.

Entrepreneurs are those managers who in one way or another were involved in the founding of their own company, regardless of their present rank. Some of these people are now presidents, others are technical managers, others are vice-presidents. What is distinctive about them, as evidenced also in other analyses (Roberts and Wainer 1971; Schein 1972, 1978), is not their present rank but the fact that they have been heavily involved in entrepreneurial activity.

Functional managers head a particular function other than a purely technical one. For example, vice presidents of finance and personnel directors are functional managers. Some titles, such as secretary, chief counsel, and treasurer, were found to be ambiguous, and we examined the entire questionnaire to determine whether a person with such a title belonged in the functional manager category.

Technical managers are those who are clearly in charge of a technical function such as basic research, research and development, engineering, or technical sales support. Excluded from this category are first-line technical supervisors, group heads, or team leaders who describe themselves essentially as senior technical people. The individual had to be at least two levels above the working technical level and had to list managerial responsibilities as part of his job. A further subclassification of this category concerned the technical area involved: engineering, computer applications, or science. Decisions of field were made mainly on the basis of self-description, corroborated where necessary by undergraduate major.

General managers are those who clearly occupy a position above functional management and who attained their positions through promotion rather than by founding their own company or joining a family business. Sample job titles are president, general manager, managing director, division president, and group vice-president.

Once the managerial roles were properly defined, the rest of the classification was simpler. The staff category encompasses first-level supervisors, team leaders, and project leaders. This group was further divided into *technologists* (also differentiated by technical field) and *business staff* (salesmen, financial analysts, and other functional specialists who are neither in management nor in a purely technical role).

Table 2.1 presents the final distribution of occupational roles that resulted from this procedure.[3] It lists in the top two sections all those who fall into the management or staff positions as just defined. There follows the education category, which offered no particular problems of classification. In the next section, which lists other professions, there are *consultants,* some of whom are operating in technical areas while others are in management consulting; architects, the only sizeable group in the sample that represents a "pure" profession; and a handful of other professionals such as doctors and lawyers. Finally, the table shows a small number of respondents who were unemployed at the time of the survey.

In analyzing this table one must be aware that the frequencies given may not reflect exactly the actual distribution of alumni in the different occupations. Even though the sample represents more than 60% of the population

Table 2.1 Basic Occupations of the Alumni Sample

Occupational Category	Number	Percentage
Management	**580**	**43%**[a]
Entrepreneur	82	6%
General Manager	50	4%
Functional Manager	157	12%
Technical Manager:		
science	20	1%
engineering	192	14%
computer applications	40	3%
other fields	4[b]	
Other Managers:		
family business	29	2%
other, not classifiable	6[b]	
Staff	**441**	**33%**[a]
Staff Technologist:		
science	60	4%
engineering	268	20%
computer applications	38	3%
other, not classifiable	5[b]	
Business Staff	70	5%
Education	**147**	**11%**[a]
University or College Professor:		
science	69	5%
engineering	51	4%
computer applications	11	1%
other fields	9	1%
Junior College, High School & Other	7	1%
Other Professions	**162**	**12%**
Consultant:		
engineering	27	2%
computer applications	4[b]	
management and other	29	2%
Architect/Planner	43	3%
Lawyer	16	1%
Doctor	21	2%
Other (e.g. minister, writer, artist)	22	2%
Unemployed	**21**	**2%**
Total	**1,351**	**100%**[a]

a. Percentages do not always add up because of rounding errors.

b. Less than 0.5%.

surveyed, there is no way of knowing whether nonresponse is systematically correlated with occupation. In particular, previous experience with response rates to a mail questionnaire of technology and chemistry alumni (Shuttleworth 1940) indicates that the number unemployed and the number employed outside the field of their training are likely to be underrepresented. One must keep this in mind when considering the occupational distribution.

As table 2.1 shows, almost half of the respondents are in some form of management, almost equally divided between technical and nontechnical areas. Another third are staff employees of organizations, primarily performing technical functions. Only 1 in 10 is in education, and the remainder are in the other professions shown. Fewer than 2% of the male respondents said they were unemployed at the time of the survey. All in all, the distribution of occupational roles is about what would be expected for a group of MIT alumni, except that the group of architects, for whom MIT provides special training, is small.

Division of this classification by year of graduation (table A.1 in appendix A) shows that the oldest two classes—those 15 or more years into their careers at the time of the survey—have very similar distributions. In the youngest class, however, just over 10 years past graduation, some differences do emerge: There are more people in education and more technical employees, and many fewer managers. The smaller percentage of managers is undoubtedly due to age and career stage. One would expect that in the ensuing 4–8 years many of these younger alumni in staff or technical positions would be promoted into management, after which the distribution of managers would approximate that in the older groups. The jump in the number of professors, on the other hand, probably reflects a real change in initial career choices based on changing national priorities and social values. Since the graduation of these alumni such social changes have led to a decided broadening of the opportunities to prepare for other kinds of careers at MIT, but in the sample of graduates from the 1950s evidence of these trends is still minimal. What strikes one more is the degree of similarity across the 1951, 1955, and 1959 classes. Here is a group of people who are fairly well established in a set of primarily managerial and technical careers. They are people who chose a particular occupational direction early in their lives and are now in the process of stabilizing it (Super et al. 1963). The education they received, the time in which they received it, and the assumptions on which their initial career choices were based are all reflected in their later occupational roles.

Eleven Occupational Categories

The classification presented in the preceding section describes the occupational roles of the sample in a way that is congruent both with accepted social categories and with the particular bent of this technically trained group. However, for the main purpose of this book, which goes well beyond description of a particular alumni group, a more compact categorization is necessary.

It is still important to keep engineers and scientists separate, even though these groups are often combined in occupational research. On the other hand, computer applications (still relatively infrequent in this group) can be combined with engineering. Similarly, it seems logical to combine all the consultants, no matter what their field, since the essence of their occupational functioning is the same: They work in small consulting firms on problems supplied to them by clients. Certain other categories, for instance the doctors, the lawyers, and the teachers below the college level, are too small for analytic purposes and thus must be eliminated. Though they probably represent interesting individual accommodations to career issues, there are too few in each role to allow analysis.

The architects present a problem. Clearly they are a well-defined group, specifically trained for their profession at MIT, and just numerous enough to permit further analysis. But they do not represent a very large proportion of the sample, and as representatives of a pure profession they are involved with issues that are different from those of most of the other groups. Had there been more of them, the desirability of including a pure profession would have called for retaining them as a separate group. But the combination of small numbers and differences in character dictated the decision to exclude the architects from the rest of this study.

Thus, we end up with eleven occupational categories which encompass the bulk of the male alumni in the sample. They are presented, along with the number and per-

Table 2.2 Eleven Occupational Categories

	Number	Percentage
Entrepreneur	82	6%
General manager	50	4%
Functional manager	157	12%
Science manager	20	1%
Engineering manager	232	17%
Staff scientist	60	4%
Staff engineer	306	23%
Business staff	70	5%
Science professor	69	5%
Engineering professor	62	5%
Consultant	60	4%
Total	1,168	86%[a]

a. These percentages do not add up to 100% because they are based on all 1,351 men in the sample, not only on those whose occupations fall into these basic categories. 140 men (10% of the sample) are unemployed or in scattered occupations that do not fit these categories, and there are 43 architects.

centage in each, in table 2.2. As the table shows, most of the alumni are still in occupations with a technical core, primarily as staff engineers and engineering managers. But there also is a not insignificant proportion in nontechnical positions, most as functional managers but some in other business functions.

In the next chapter it will become clear that these eleven occupational roles may themselves be grouped into two basic career patterns. But first, let us see if there are differences among the people occupying them in current employment and family characteristics or in socioeconomic origins. (The data are presented in table A.2, appendix A.)

Though none of the differences with respect to occupational status of the respondents' fathers is very large, a number of things are worth noting. Among alumni who are entrepreneurs there is a disproportionate number of businessman fathers, particularly in small businesses, whereas among the general managers one sees a disproportionate number of fathers in major business enterprises. Very few of the general managers come from blue-collar homes. It is the engineering managers, staff engineers, and science professors who are most likely to have had fathers with blue-collar occupations. (The science professors, it must be said, also have a high proportion of fathers in the top professions.) All in all, as other occupational research has shown (Osipow 1973), father's occupation is somewhat correlated with son's occupation. As is often claimed, there is also an indication that engineering as a profession provides relatively easy social mobility. Finally, science seems to attract children from both ends of the socioeconomic scale. It is probably the field where ability counts for the most in guiding people into particular career paths.

Two current employment characteristics show differences among the roles. The first is the sector of the economy within which the alumni are now working. As might be expected, almost all of the alumni now in nontechnical roles work in the private sector. The professors, of course, are all in nonprofit institutions. But the differences between science and engineering managers, and between staff scientists and staff engineers, are of substantive interest—the staff scientists and the science managers are more likely to work in nonprofit or government laboratories than are their engineering counterparts. As will emerge in the next chapter, these groups fall into different career patterns. Second are differences in income. The general managers and entrepreneurs earn the most, followed by the consultants. Alumni in faculty and staff positions have the lowest incomes. There is not much difference between the salaries of engineers and scientists except between the professors in those fields, but this difference turns out to be a function of the younger age of the science professors. (Other differences in income do not disappear when one controls for age; see Bailyn and Schein 1972.) It is also of interest that the incomes of the engineering professors, who have more advanced degrees, are not very much higher than those of the staff engineers and are lower than those of the engineering managers.

There are also some differences among the roles in stage of family, wife's professional status (based mainly on level of education), and wife's working status. Not surprisingly, the managerial groups are further along in their family cycle than most other groups and the professors are less far along; this reflects the relative ages of these groups. The managers are most likely to have nonprofessionals as wives (the extreme case of this is the general managers), whereas the wives of professors and staff scientists are more likely to be professionals. The data with respect to whether or not wives are currently working show similar results. The general managers are the least likely and the professors and staff scientists are the most likely to have working wives. These differences are important because they are related to the alumni's reactions to their work and to their families.

The general managers stand out on a number of characteristics. They are most likely to have had fathers in major business roles and least likely to have had fathers in occupations lower on the socioeconomic scale. They have by far the highest incomes, and their wives are clearly in nonworking support roles. As has been suggested in a previous paper, they may well represent a group of people who made an early choice of management as an occupation on the basis of a combination of socioeconomic and value factors (Schein 1972).

In general, however, it was the educational and first-job experiences of the alumni that propelled them along certain career paths. And when one looks at those experiences, one sees a distinct pattern among these eleven occupational roles.

Notes to Chapter 2

1. When they entered the work force, 51% went into basic engineering positions and another 12% combined engineering with other duties. These figures are based on the 87% of the respondents for whom information on first jobs is available.

2. This definition of "staff" differs somewhat from the use of "staff" as distinct from "line" (Pigors and Myers 1977), which is based primarily on functional distinctions.

3. Coding into occupational categories was done by the two researchers in conjunction with Dany Siler, research assistant to the project at that time. Error was minimized by having at least two of us independently code each questionnaire. In about 95% of the cases (based on a check of one-fourth of the questionnaires) the first two coders agreed on their classification, which then became final. If disagreement remained after discussion between the two initial coders, the questionnaire was given to the third person and the final classification was based on the consensus of all three researchers.

3 Career Patterns and Values

Occupational research often seeks to classify occupations into a small number of broad categories in order to explore the individual characteristics that best fit with certain general occupational requirements. Various criteria have been used to arrive at such classifications, as is evident from the work of vocational psychologists (see, for example, Strong 1943; Roe 1956; Kuder 1957; Holland 1959, 1966). With the relatively homogeneous group of respondents in this study it was possible to group the eleven occupational roles defined in the preceding chapter into two categories. These two patterns emerged from an analysis of specific, objective career data. Subsequent analysis of more subjective perceptions and attitudes showed that the two patterns are also associated with two distinct value syndromes. This validation of the classification by differences in value norms gives some assurance that these career patterns reflect the actuality of people's lives and are not only external taxonomic categories.

The Two Major Career Patterns

Table 3.1 shows two basic clusters of occupations which are more or less homogeneous with respect to undergraduate major, academic performance as an undergraduate, graduate school attendance, and first job. One cluster, which we have called the *engineering-based* careers, pattern E, includes the entrepreneurs, the general and the functional managers, business staff, the consultants, the engineering managers, and the staff engineers. These are the various roles in which an engineer may find himself some 10–20 years after his engineering training. The alumni whose careers fall into this pattern tend to be graduates of the School of Engineering, to have obtained no more than average grades, not to have doctorates, and to have started their careers in engineering staff positions. While there are some variations within the cluster (such as the obvious contrasts between general managers and staff engineers), what is more striking is the degree to which people in these careers had similar early-career experiences. The other major cluster—the *scientific-professional* careers, or pattern SP—shows a different history. These alumni (the group includes the science professors, the engineering professors, the science managers, and the

Table 3.1 Career Patterns Based on Early-Career Events

	Pattern E: Engineering-Based			
	Total 957[a]	Entr. (N=82)	Gen. Mgr. (N=50)	Funct. Mgr. (N=157)
% from school of				
Engineering	77%	70%	64%	63%
Science	6%	11%	4%	4%
Management	15%	18%	32%	27%
% with honors[b]	19%	19%	29%	16%
% with doctorate	11%	8%	24%	6%
% whose first jobs were in engineering staff positions	61%	54%	48%	51%

a. These N's are reduced by the number about whom information on any category is not available.
b. Overall grade average B+ or higher.

staff scientists) more frequently graduated from the School of Science (with the exception of the engineering professors), obtained higher grades, were very likely to have gone on for a doctorate, and were less likely to have started in engineering jobs.

Of the total alumni sample 71% fall into pattern E. Of these, 56% are still in clearly technical jobs, such as engineering manager or staff engineer, while 9% have become entrepreneurs and the remaining 35% have gone into various other business roles or into consulting. Only 16% of the male alumni fall into the SP pattern. Almost 10% of these are managers, but they are a special kind of manager distinguished both from the business-oriented and the engineering-oriented managers. They are more formally educated and are associated with more science-oriented organizational settings. They work in universities, for the government (particularly the military), or in industrial research and development. Most have doctorates and most identify themselves as scientists. But even those who identify more with management or who stopped short of the doctorate definitely had science-oriented positions, such as assistant director of a military solid-state physics research lab, research manager of R&D in the chemical industry, or manager of an environmental and applied science center. Only one alumnus in this group has neither a

				Pattern SP: Scientific-Professional				
Bus. St. (N=70)	Cons. (N=60)	Eng. Mgr. (N=232)	St. Eng. (N=306)	Total 211	Eng. Prof. (N=62)	Sci. Prof. (N=69)	Sci. Mgr. (N=20)	St. Sci. (N=60)
66%	60%	85%	87%	34%	74%	7%	30%	23%
6%	8%	6%	7%	60%	6%	90%	70%	77%
27%	27%	6%	6%	3%	10%	—	—	—
16%	23%	20%	19%	47%	64%	46%	31%	35%
6%	12%	13%	11%	88%	98%	86%	80%	72%
59%	46%	57%	77%	21%	33%	12%	22%	18%

doctorate nor a self-identification as a scientist, but at the time of the survey he was on a temporary assignment as chief scientist in a branch of the military.

At the staff level, too, the scientists and engineers fall into different patterns. There is a different degree of professionalization in the two groups, as indicated primarily by the high proportion of doctorates among the staff scientists and the very low proportion among the staff engineers.

In contrast, the faculty members in the two fields (and over 60% of the alumni in the SP pattern were in academic careers at the time of the survey) fall into the same cluster. The engineering professors fit into the SP pattern by every criterion except undergraduate major. What this seems to imply is that in engineering an academic career represents quite a different path from a staff or a managerial one, whereas in science the academic, staff, and managerial roles are more interchangeable.[1] Further, this branching of the academic career in engineering from the other engineering paths seems to occur in college. The engineering professors had the highest grades of any group in the sample, which suggests that high academic performance as an undergraduate acted as a powerful force toward entering graduate school, obtaining a doctorate, and ultimately selecting an academic career. Two engineering graduates, both now engineering managers in the public sector, comment on this tendency. The first is disappointed that MIT discouraged him from going to graduate school:

> There was a ridiculous misconception perpetrated, that anyone with less than a 4.0 [grade point average] had no business considering graduate school. They dashed and trampled individual self-confidence at every turn.

The other, with a higher grade point average, did go on to get his doctorate but now says:

> My undergraduate education was too much a preparation for graduate school and my graduate education prepared me for a career at a university but not in the world.

All in all, the engineering-based career pattern is more varied than the scientific-professional pattern because it not only leads to technical engineering positions but is also the primary source of the business-oriented careers in this sample. In contrast, the SP pattern leads to a more circumscribed set of occupational roles, primarily as research scientists and college and university faculty.

Differences in Job Values

The two major career patterns emerged from an examination of differences in external or objective indicators, particularly those surrounding early-career events. But though these differences are clearly evident, they do not reveal whether these patterns are experienced differently by the people in them. If career patterns are to be a useful way of understanding people's work lives, they must not only represent differences in external characteristics but must also have an internal reality. It is important, therefore, to investigate other differences between and within these patterns to see if there are internal indicators that parallel the differences already described.

The aspects of a job that a person considers particularly important are an indication of the internal needs he would like to have met by his work. A number of previous studies have tried to identify occupational differences in job values, and have found such differences even between closely aligned occupations such as scientists and engineers. For instance, autonomy—the freedom to follow one's own ideas—is especially important for the scientist, who seeks a job that allows him to pursue his own interests (Glaser 1964, Pelz and Andrews 1966; Reiss and Balderston 1966). The scientist has also been shown to be particularly likely to have "intellectual" values: to want a job where he can be creative, select his own research areas, and work with ideas (David 1971). The engineer is more likely to be concerned with moving up the organizational ladder and seeing his ideas put to use in his organization (Marquis 1965; Perrucci and Gerstl 1969). He is concerned less with intellectual values and more with external factors surrounding his work (Rosenberg 1957; David 1971).

Results from the present study also show that job values differentiate among technical careers. The data are based on the following question (item 17 of questionnaire):

The list below shows a number of characteristics of a job. Please circle the appropriate number to show how important you feel each characteristic is to you with regard to your present and future jobs.

The response scale ranges from 1 ("not at all important") to 5 ("very important"). The list consists of 22 job characteristics, and includes highly intrinsic values such as "challenging work to do" and highly extrinsic or instrumental ones like "good physical working conditions" or "good fringe benefits" as well as items concerned with the value placed on working with people and being concerned with society.[2] All the items are presented in table 3.2, in the order in which they were deemed "very important" by the total sample.[3] It is obvious from the table that those job characteristics considered most important in this sample concern the actual work, whereas those viewed as least important are concerned with more peripheral aspects of the job.[4] This finding, indicating the expectation that internal needs of competence and achievement can be met by the work one does, corroborates the view already set forth about the men in this sample and the importance they attach to the careers they are pursuing.

But the focus here is not the job values held by the sample as a whole. Rather, the emphasis is on the differences in values between the two career patterns. Do engineering-based careers conform to different needs and values than scientific-professional careers? Or are there more significant variations in job values among occupations within a career pattern? In attempting to answer such questions on the basis of individual responses, one faces an immediate problem: Some people find themselves in

Table 3.2 List of Job Values, with Percentage Rating Each "Very Important"

Work from which I could get a personal sense of accomplishment	70%
Challenging work to do	66%
Considerable freedom to adopt my own approach to the job—to be creative and original	52%
Opportunity for advancement	48%
Job that allows me to make a real contribution to the success of the organization	45%
Opportunity to exercise leadership	33%
Recognition for doing a good job	30%
Opportunity for high earnings	27%
Job that leaves sufficient time for family and personal life	21%
Location	20%
Department where people are friendly and congenial	17%
Reasonable work load	15%
Job that allows me to make a contribution to society	14%
Highly regarded organization	13%
Opportunity to work with people rather than things	12%
Job highly regarded by others in company—a job with some prestige	11%
Training or educational opportunities (to improve my knowledge and skills)	11%
Efficiently run department	10%
Job security	10%
Work that is relevant to social problems	5%
Good fringe benefits	5%
Good physical working conditions	5%

occupational roles that do not permit fulfillment of their job-related needs and values. The characteristics they consider most important about a job may not be found in their current work, and therefore are not indicative of occupational norms. Since the concern in this chapter is with occupational roles rather than with individual responses to work, one would like to eliminate cases of such lack of fit as much as possible. There is no way with the data at hand to isolate exactly the individuals whose needs or values fit their occupational roles, but one can approximate such a fit by limiting the analysis to those people in each occupational group who consider themselves successful.[5] Values expressed by such a group —where the fit between needs and occupational requirements is likely to be high—are indicative of general occupational norms and thus provide information about the patterning of careers on the basis of internal indicators.

Table 3.3 gives the percentage of the "successful" [6] men in each occupational group and in each pattern who consider a particular job characteristic "very important." The items shown in the table are limited to those that differentiate the engineering-based from the scientific-professional career pattern—those in which there is an overall difference between the two patterns that is not dependent only on a particularly high or low percentage in a single occupational role but is reflected in a majority of the occupations within a career pattern.[7] The upper four job characteristics listed in table 3.3 are more highly valued by "successful" alumni whose occupations fall into the engineering pattern than by those in the scientific-professional one. These four, which make up what one might call the "engineering syndrome," are opportunity for high earnings, opportunity for advancement, opportunity to exercise leadership, and possibility of making a real contribution to the success of the organization. With few exceptions, every engineering-based occupation rates these items higher than do the scientific-professional roles. Even the staff engineers, who rate these items relatively low, place them higher than do their counterparts, the staff scientists. This syndrome is clearly associated with the values of management, and reflects an orientation toward getting ahead within an organization rather than as a professional contributor. It is not surprising, therefore, that the science managers, among the

Table 3.3 Percentage of "Successful" Respondents Rating Given Job Characteristics "Very Important"

	Pattern E: Engineering-Based			
	Total 314[a]	Entr. (N=42)	Gen. Mgr. (N=39)	Funct. Mgr. (N=53)
Opportunity for high earnings	47%	48%	65%	53%
Opportunity for advancement	68%	67%	71%	77%
Opportunity for leadership	57%	65%	74%	70%
Contribution to organization	60%	60%	82%	72%
Freedom to be creative and original	56%	68%	63%	53%
Educational opportunity	13%	10%	14%	11%
Sense of accomplishment	77%	74%	82%	77%
Challenging work	76%	78%	87%	74%

a. N values represent the number of people in each category who consider themselves "successful," reduced where necessary by the number not answering a given item.

SP groups, rate most of these values highest. It is also of interest that two of these values, advancement and leadership, differentiate between science and engineering professors. The fact that the engineering professors consider both these characteristics more important may reflect their initial occupational choice and expectations.

The job characteristics more valued by "successful" alumni in the scientific-professional pattern, as shown in the bottom half of table 3.3, are freedom to be creative and original, challenging work, work that gives a personal sense of accomplishment, and opportunity for further education. With the possible exception of educational opportunity, these values relate to the intrinsic characteristics of work. It is this emphasis that forms the core of the "scientific-professional syndrome." The interpretation of this syndrome is complicated by the fact that though these intrinsic characteristics are more valued by people in SP occupations than they are by those in E occupations, the latter also give high ratings to these job values. In fact,

				Pattern SP: Scientific-Professional				
Eng. Mgr. (N=75)	Cons. (N=34)	Bus. St. (N=20)	St. Eng. (N=51)	Total 71	Sci. Mgr. (N=12)	Eng. Prof. (N=25)	Sci. Prof. (N=17)	St. Sci. (N=17)
36%	52%	55%	35%	13%	17%	4%	12%	24%
75%	67%	80%	41%	35%	42%	44%	24%	29%
53%	61%	45%	33%	25%	42%	39%	6%	12%
49%	61%	75%	43%	39%	58%	33%	35%	35%
48%	61%	55%	51%	86%	54%	92%	88%	94%
11%	21%	10%	16%	25%	0	39%	24%	24%
75%	82%	65%	80%	86%	83%	83%	88%	88%
68%	88%	70%	74%	84%	75%	83%	88%	88%

except for educational opportunities, these characteristics are the most valued in the entire alumni sample, and two of them—challenging work and sense of accomplishment—are among the three highest values in every occupational group except business staff, in which they rank third and fourth. In calling this set a scientific-professional syndrome, therefore, we are not asserting that the items are valued highly only in the SP group, but that they are rated higher in this group than in the E group. Educational opportunity, though not highly rated in an absolute sense, does differentiate the E and SP patterns, with the exception of the science managers, who place very little value on it. All the other SP occupations value further educational opportunities more than the E occupations do, which probably reflects their greater initial dependence on postgraduate education.

Overall, the data in table 3.3 reveal two different value syndromes associated with the two career patterns previously identified. While all occupational groups value the intrinsic job factors such as challenge and accomplishment, these values, along with creativity and further education, clearly are more important in the scientific-professional occupations. The engineering-based occupations also value the intrinsic job factors, but, relative to the SP pattern, are much more concerned with leadership, advancement, contributing to the success of their organization, and high earnings. The SP occupations, one might say, follow "professional" norms: The work they entail is seen as intrinsically satisfying and presumably meets internal needs of competence and autonomy. The E syndrome, with its emphasis on "getting ahead through organizational routes," corresponds more closely to the norms associated with the successful "organizational" career.

Socialization or Preselection?

With cross-sectional data it is difficult to determine whether such value differences are the result of individual attitudes and traits that *determine* career choice ("preselection") or whether they develop *in response to* occupational experiences ("socialization"). It seems most likely that both processes are at work. In studies of college students and their occupational choices, both directions of influence have been adduced. Rosenberg (1957) concluded that values have a greater effect on occupational choice than choice has on values; Underhill (1967) felt that it is more likely that the career itself leads to "appropriate occupational values." Athanasiou (1971), in a study of college students who had initially opted for an engineering major and then changed their minds, found that students who transferred had preexisting differences from an equivalent group of constant engineering majors, even before the freshman year; however, the fact that the differences between the two groups were considerably greater in the middle of the sophomore year, after the decision whether to stay in engineering had been made, showed that "career" influences had accentuated the existing differences. In a panel study of M.B.A.s, Schein (1978) also found evidence for both preselection and socialization. Similar findings for differences in cognitive style were reported by Kolb and Fry (1974).

It seems likely, therefore, that the same holds true for the alumni studied here. Certain differences in values and personal traits were no doubt already present while these men were still in school. But since MIT alumni can be assumed to have had many similar predispositions at the point of entry into their careers, it becomes highly probable that part of the variation found results from career-induced changes. Though it is not possible to prove this point with the data at hand, one can examine the possible existence of occupational socialization by comparing, within the engineering-based career pattern, alumni with doctorates to those without them. The logic of this analysis rests on the assumption that at the point of entry into their careers, people with doctorates are more likely to adhere to the scientific-professional values than are people without such advanced degrees.[8] The following question then arises: What happens to those values in career patterns or occupations that do not reinforce them strongly? More specifically, what happens to the holder of a doctorate who successfully enters an occupation in the E pattern? [9] If doctorate holders in E-pattern occupations resemble in their values the SP alumni, one can argue that preselection makes the difference. In such a case it would be the doctorate, the common element in the two groups, that would seem to explain the value norms. If, in contrast, these E-pattern Ph.D.s or Sc.D.s resemble more the other E-pattern alumni, one would have some assurance that occupational roles have modified their values (though it is also conceivable that this group for some reason had never "fitted" the doctoral mold in the first place).

The results in table 3.4 show evidence for occupational socialization. The first column presents data from alumni who do not have a doctorate and who are in E-type occupations (occupations congruent with their educational level). The third column represents alumni who have doctorates and are in "congruent" SP occupations. It is the middle column that permits some test of the socialization hypothesis. It represents alumni who have doctorates but who are now in "incongruent," E-type occupations. If their values differ from those of doctoral-degree holders in SP occupations and more closely resemble those of the non-doctoral alumni in E occupations, then there is evidence for the effect of the career on the individual.

It is clear from the table that such an effect is present. For the values of earnings, leadership, creativity, educational opportunity, and challenge the middle group is almost identical with the rest of the E pattern. On the other

Table 3.4 Percentage of "Successful" Alumni in Congruent and Incongruent Occupations Rating Given Job Characteristics "Very Important"

	Pattern E		Pattern SP
	No doctorate; "congruent" occupations (N=277)[a]	Doctorate; "incongruent" occupations (N=37)[a]	Doctorate; "congruent" occupations (N=61)[a]
E-Syndrome Values			
Earnings	47%	46%	5%
Advancement	69%	57%	36%
Leadership	57%	57%	22%
Contribution to organization	62%	51%	35%
Average percentage	59%	53%	24%
SP-Syndrome Values			
Being creative and original	56%	51%	85%
Educational opportunities	13%	11%	25%
Feeling of accomplishment	77%	81%	87%
Challenging work	76%	73%	85%
Average percentage	56%	54%	70%

a. N values represent the number of people in each category who consider themselves "successful" in their work. They are reduced, where necessary, by the number not answering the relevant questions.

items, the middle group has a position between the first and the third (though consistently closer to the first), which indicates both preselection and occupational socialization, with the latter effect stronger. This conclusion is confirmed by the average percentages given in the table. To a great extent, therefore, the value differences between the engineering and the scientific-professional career patterns appear to result from career experiences, and not only from initial differences between the people who enter these fields.[10]

Summary

This chapter has shown that a large proportion of the careers of the male MIT alumni fall into two basic career patterns. Over 86% of the total sample are in engineering-based or scientific-professional careers. The work entailed in these careers flows fairly directly from the technical training of the respondents and from relatively homogeneous sets of early experiences. These beginnings have shaped the ensuing career paths; each of the career patterns leads to a fairly constrained set of occupational roles. Successful movement along these paths seems to introduce or accentuate an associated group of values. At mid-career, therefore, successful careers in the engineering pattern have come to be associated with a set of values different from those of successful scientific-professional careers.

Occupations in the engineering pattern—whether technically oriented or more generally managerial—inculcate organizational values of leadership, advancement, and earnings, which are superimposed on the more intrinsic values of opportunities for job challenge, sense of accomplishment, and freedom to be creative. Scientific-professional occupations, in contrast, show no such blending. They are geared much more to the intrinsic "professional" values, and much less to the organizational ones.

Notes to Chapter 3

1. As further corroboration of this difference, we note that there is much more movement into teaching positions from companies or labs among scientists than among engineers. If we look at all people whose first jobs were in engineering staff positions (no matter what career pattern they now fall in), only 5% are now teaching. If, in contrast, we take all those whose first jobs were in science staff positions, 29% are now teaching in some capacity.

2. The list has been adapted from similar ones used in studying occupational values of students (Rosenberg 1957; Davis 1965) and in surveying sales, service, and technical personnel in the IBM World Trade Corporation (Sirota and Greenwood 1971). A number of previous studies have grouped such items into three similar areas. Rosenberg (1957) determined three major "value-orientations" or "value foci": a "people-oriented value complex," an "extrinsic reward-oriented value complex," and a "self-expression oriented value complex"; Davis (1965) identified three independent dimensions of values: "people," "money," and "original and creative"; David (1971) defined three "values indexes" to describe the occupational values of scientists and engineers: a "people-orientation index," an "external factors index," and an "intellectualism index."

3. Only if a person rates a given characteristic as "very important" does it make sense to say that it represents an internal need he would like met by his work. It is for this reason that the

percentage of a given group rating a particular characteristic "5" is used as a measure of job values.

4. Work that relates to society and its problems is rated relatively low in this group. One would expect more recent classes to be more socially concerned, but there is some question whether this greater concern persists through the midcareer years.

5. A discussion of the definition and measurement of occupational success and the occupational differences on these success measures is found in appendix C.

6. This includes all respondents who when asked "At this point in your professional life, how successful do you think you are in your work?" rated themselves "very successful" ("5" on a 5-point scale) and those who rated themselves as medium ("4") in success at that time but indicated on a later question that they expected to be "very successful" ("5") at the height of their career.

7. The responses to those items not included in table 3.3 are given in table A.3, appendix A.

8. Whether this difference at entry results from educational socialization or from already existing differences between students who seek doctorates and those who do not is unknown but not crucial for the attempt to investigate the possible effects of occupational socialization later in the career.

9. The logically equivalent question of what happens to the person without the doctorate who enters an SP occupation is difficult to test, since there are only 10 alumni without doctorates who view themselves as successful in SP occupations (and only 25 in all).

10. The data on the 10 alumni without doctorates now "successfully" engaged in SP occupations, though not very reliable because of the small N, indicate that this minority has almost the same values as the predominant SP group on those items that define the SP syndrome, where their average is 68% (indicative again of occupational socialization), but shows some deviance in the E syndrome, where the average is 48%. On these organizational values they have an intermediate position, somewhat closer, in fact, to the engineering-based occupations. Closer analysis shows that on these organizational values they exhibit a bimodal distribution: Some deem them quite unimportant (as expected in SP occupations), but some view them as very important (more in line with their lack of doctorates than with their occupational roles).

4 Reactions to Work

The "leading positions" for which MIT trained its graduates have been viewed, until quite recently, as among the most desirable in our society. True, these roles demanded high commitment of time and effort, but the rewards were great. The positions conferred status and power and—perhaps more important—satisfied the needs of the scientists and engineers to express competence, to master some portion of the environment, and to continue to grow. All of the respondents in this study, presumably, had embarked on their careers with the expectation that the work they would be doing would be intrinsically satisfying—an expectation succinctly expressed by one alumnus who added to the list of important characteristics of a job one that was central for him: "fun (in my work)."

By considering the reactions to work of the respondents this chapter deals with the midcareer outcomes of this expectation. As will be seen, there are a number who are disappointed, whose expectations are not being met. Forces stemming from the particular turns their careers have taken and from changes in social values have led in these cases to problems similar to those confronting the "alienated" industrial worker (Blauner 1964).

The primary analytic question raised in this chapter is whether such differences in reactions to work can be systematically related to occupation. Though there are some identifiable differences in these reactions between alumni in the engineering-based and those in the scientific-professional career patterns, there also are significant differences within each pattern. The relative homogeneity of these patterns seems to break down when one looks beyond group norms and values and shifts the focus of the analysis to the way individual alumni actually react to their work in the middle years. It is not necessary, however, to revert to the eleven occupational categories of chapter 2. Rather, it turns out that a simple distinction within each of the two career patterns, based on the level of the current organizational position, is sufficient. In other words, though the reactions to work of this sample of technically trained alumni cannot be systematically related to the two career patterns identified in the preceding chapter, they can be understood in terms of four broad categories.

Work Involvement

The questionnaire was designed to assess reactions to work in as full a way as possible. It thus contained many items dealing with the satisfactions derived from job, work, and career, with important job values, with self-assessments of success and other traits and skills, and with the role of work in a person's total life. Only such a large set of items, we felt, would make it possible to capture the complexity of these respondents' work situations and of their reactions to them.

From the analysis of the interrelation of all of these items emerged one key concept, which we call work involvement. The exact derivation of the index of work involvement, as well as the details of attempts at explication and construct validation, are given in appendix D. The final index consists of items that pertain to commitment to and satisfaction derived from professional work and to the relative weight of career and family in overall life satisfaction and in job decisions.[1]

It is the inclusion of items centering on the relative role of career and family that perhaps more than anything else distinguishes this definition from previous measures of work involvement or commitment to work. Though there is no logical incompatibility in combining or integrating high commitment to work and great involvement with family, in this sample (perhaps because of the particular range of occupations represented) such integration is not usual. It is primarily the managers who seem able to respond to the demands from their families as well as to those from their work. However, since relatively few of the managers' wives are involved in their own work, the manager's ability to combine commitments in these two life areas depends, apparently, on his wife's playing an ancillary role to his career. This is very different from what is envisioned in a "dual-career" family, a family in which the wife's role is not defined solely by what her husband does. Results of studies on the success of this pattern (Bailyn 1970; Holmstrom 1972; Rapoport and Rapoport 1976) point to the need for the husband in such a family to withdraw somewhat from the fully work-committed stance traditionally expected of professionals in our society. Even the managers, therefore, do not contradict the tendency in this sample for commitment to work and commitment to family to be negatively associated.

In general, respondents who score high on the index of work involvement developed in this study derive some of their main satisfactions from their work, and it plays the major role in their lives—above family and above creative, recreational, or community pursuits. Individual analysis of the questionnaires of the 25 alumni with the very highest work-involvement scores shows that they know exactly where they are professionally and what they want to get from their jobs. They have a very active orientation to their work, and if their particular jobs do not satisfy them they tend to make plans to change. For these highly involved respondents, intrinsic needs for the expression of competence and for mastery and achievement are being fulfilled by the work they do.

At the other extreme, those who are low on the index of work involvement approach their work more instrumentally and derive their major satisfactions from their families or from other non-work-related sources. They do not find their work rewarding in its own right; for them it is more a necessity than a pleasure. Reading over the questionnaires of the 23 people with the very lowest scores on the index, one gets the impression that the concern with work of this group is with its financial rewards or with the fact that it gives them something to do with their days. They seem to lack self-confidence, and a few are very bitter. As one wrote, "the myth of a 'shortage of engineers' and an unfortunate scholarship to MIT *seduced* me from a medical or legal profession." Quite a few of them express the desire to have followed a different profession, and yet they show no evidence of taking initiative as far as their jobs are concerned. Many are still with the same organizations they have always been with, and hardly any are making plans to change their working situations. They seem to have no firm professional identity, and give evidence of having ill-defined, unrealistic work goals.

Table 4.1 presents the distribution on the index of work involvement. It shows, for each of the occupational roles in the two major career patterns, the percentage of respondents whose work involvement is high, medium, and low. Further, it presents the mean work-involvement score for each group and the extent to which the group is

Table 4.1 Work Involvement by Occupational Role

		Work Involvement[b]				
	N[a]	% High	% Med.	% Low	Mean Score	D (% High − % Low)
Engineering-Based Pattern						
Entrepreneurs	73	33%	54%	13%	3.17	+20
General managers	47	29%	52%	19%	3.09	+10
Functional managers	151	24%	58%	18%	3.07	+6
Consultants	58	26%	53%	21%	3.07	+5
Engineering managers	226	28%	47%	25%	3.02	+3
Staff engineers	286	20%	44%	36%	2.77	−16
Business staff	68	9%	62%	29%	2.77	−20
Total	909	24%	50%	26%	2.96	−2
Scientific-Professional Pattern						
Science managers	19	48%	42%	10%	3.28	+38
Science professors	66	40%	50%	10%	3.34	+30
Engineering professors	57	33%	57%	10%	3.27	+23
Staff scientists	56	16%	68%	16%	3.06	0
Total	198	32%	56%	12%	3.24	+20

a. Number of people who answered all the questions in the work-involvement index.
b. On a scale from 1.0 to 5.0, scores $\geq$ 3.6 were considered high and scores $\leq$ 2.0 were considered low. These cutoff points were chosen in order to differentiate the high and low groups clearly and to equalize as much as possible the percentages in the two extreme groups for the sample as a whole.

more characterized by people with high than with low work involvement. (The latter measure, *D*, determines the order in which the occupations are presented within each of the career patterns.)

Organizational Position

As table 4.1 shows, the occupational roles with the highest degree of work involvement (those in which there are many more highly involved than uninvolved respondents) are in the scientific-professional pattern: science managers and science professors are the most involved groups, followed by engineering professors. At the other extreme, the table shows that alumni in two occupations in the engineering-based pattern (staff engineers and business staff) are least involved in their work. In general, as the figures in the "total" rows of the table indicate, the engineering-based careers are less involving than the scientific-professional ones.

But this general statement hides some sizable and systematic differences within each of the career patterns. In particular, the staff positions in business, engineering, and science show considerably less work involvement than do the other positions listed. It is this result that leads to the further distinction of occupational roles within the two career patterns on the basis of *organizational position*. One of the two types of organizational positions, then, consists of staff roles, which are the roles whose organizational level is likely to be "low." The other type of position is somewhat more complex. It includes managers, consultants, entrepreneurs, and university faculty. The common thread among these might be prestige, or possibly income, or the extent to which the position permits autonomous functioning in the work role. We will call these occupational roles "high" in organizational position, as they are likely to be higher than the staff roles in all these ways.[2]

The distinction between high and low organizational positions is associated with a characteristic career attitude. As table 4.2 shows, occupants of high organizational positions are more likely to consider success at work very important and to have high aspirations for their careers than are those whose organizational positions are low. Whether this association comes about because success aspirations are a prerequisite for high organizational posi-

Table 4.2 Success Aspirations of Occupants of High and Low Organizational Positions

		Success Aspiration[b]		
Organizational Position	N[a]	High	Medium	Low
High	717	34%	31%	35%
Low	421	15%	25%	60%

a. Number who answered both questions cited in the following note.

b. Measured by the responses, on an ascending scale, to the questions "How important is it to you to be successful in your work?" (question 20) and "[To what degree do you now possess] high aspirations for your career?" (part of question 29). All those who said that success at work is very important to them ("5") and *also* indicated that they possess high career aspirations to a great extent ("4") are considered to have high success aspirations; those who met neither of these criteria are considered low in success aspirations.

tions or because these positions inculcate such attitudes cannot be said with certainty, but, following the reasoning about values in chapter 3, we presume that both processes occur and that the effects are reciprocal. People who pursue or are selected for high organizational positions may be somewhat inclined toward high success aspirations in any case, but this stance is undoubtedly accentuated and strongly reinforced by the way these positions are experienced by their occupants.

Four Types of Technology-Based Careers

The investigation of *individual* reactions to work leads to a distinction between high and low organizational positions within each of the career patterns. Table 4.3 presents the occupational roles included in the E pattern with high positions and with low positions, and in the SP pattern with high and low positions.

The remainder of this chapter details the different reactions to work associated with these four career types. As an introduction to this discussion we present the comments of a representative alumnus from each of the four career types on his MIT education. (None of these respondents, it should be said, was a top student during his undergraduate years.) A general manager (career type E/H) writes the following:

Without a doubt my attendance at MIT and the experiences there have had the most positive and beneficial

Table 4.3 Four Career Types

	Career Pattern	
Organizational Position	Engineering-Based (E)	Scientific-Professional (SP)
High (H)	**Type E/H (N=581)**	**Type SP/H (N=151)**
	Entrepreneur (N=82)	Science manager (N=20)
	General manager (N=50)	Science professor (N=69)
	Functional manager (N=157)	Engineering professor (N=62)
	Consultant (N=60)	
	Engineering manager (N=232)	
Low (L)	**Type E/L (N=376)**	**Type SP/L (N=60)**
	Staff engineer (N=306)	Staff scientist (N=60)
	Business staff (N=70)	

effects on my life other than the upbringing given me by my parents. MIT served to broaden my outlook on life and expose me to a wide variety of professional opportunities about which I would have never known. The education has been a cornerstone to my everyday life. It has affected my career, influenced my marriage, and enabled me to undertake an ambitious career.

The alumnus in type SP/H, a professor of physics, reports that

MIT was where I did most of my growing up socially, professionally, and spiritually. MIT was my home, my school, and a way of life which I shall always cherish.

The alumnus in type E/L, a development engineer, has a different kind of memory:

My training at MIT was "thing" as opposed to "people" oriented. The school operated as an efficient but a cold business organization. I feel that I graduated as a well trained and competent technician but lacked a feel for the perspective of my profession. I learned a trade but what I was doing always appeared to be out of context.

And an alumnus now in a science staff position (type SP/L) complains that he had

difficulty in understanding what the total picture of the working scientist was. All I was presented was rather specific courses in the "pieces," i.e. couldn't see the forest for the trees.

It certainly seems more reasonable to attribute such dramatic differences to the later career than to blame (or congratulate) the institution in which each of these individuals was educated.

Engineering-based careers, as shown in chapter 3, tend to be governed by organizational values, scientific-professional ones by professional values.[3] And, as shown at the beginning of the present chapter, alumni in high organizational positions tend to have higher success aspirations than do those in low positions.[4] Thus, career patterns—independent of organizational position—are associated with certain value syndromes, and organizational position is independently related to a characteristic stance toward success and career. These differences are summarized in table 4.4, which also gives the work involvement in each of the career types.

It is clear from this table that work involvement relates

Table 4.4 Occupational Value Norms and Characteristic Career Stance of Four Career Types (Including Data on Work Involvement)

Organizational Position	Career Pattern: Engineering-Based (E)	Career Pattern: Scientific-Professional (SP)
High (H)	**Type E/H (N=555)**[a] Organizational value syndrome. High concern with success and career. Work Involvement[b] H 27% M 52% L 21% $D = +6$	**Type SP/H (N=142)**[a] Professional value syndrome. High concern with success and career. Work Involvement[b] H <u>38%</u> M 51% L <u>11%</u> $D = +27$
Low (L)	**Type E/L (N=354)**[a] Organizational value syndrome. Low concern with success and career. Work Involvement[b] H <u>18%</u> M 47% L <u>35%</u> (double underlined) $D = -17$	**Type SP/L (N=56)**[a] Professional value syndrome. Low concern with success and career. Work Involvement[b] H 16% M 68% L 16% $D = 0$

a. N = number of people who answered all the questions in the work involvement index.

b. The χ^2 for the 4 × 3 table (4 career types, 3 levels of work involvement) is very high ($\chi^2 = 57.3$, $P < 0.001$). Partitioning χ^2 to locate the main source of significant differences is difficult because of the large differences in N's in the four career types. But analysis of residuals shows that the highest deviation (double underlining) occurs in the E/L career type, who are much more likely to have low work involvement than expected (adjusted residual = 6.09). Other significant residuals (single underlinings) occur in the SP/H group, who are more highly work-involved and less uninvolved than expected (adjusted residuals = 3.88 and 3.92, respectively), and in the E/L group, who are also less highly involved than expected (adjusted residual = 3.76) (Haberman 1973, quoted in Reynolds 1977, p. 12).

both to career pattern and to organizational position. Scientific-professionals are in general more involved than are occupants of engineering-based careers, and alumni in occupational roles representing high organizational positions are more involved than are those in roles whose organizational positions are more likely to be low. Work involvement is highest, then, in scientific-professional roles with high organizational positions—particularly (see table 4.1) among the science managers, but also among the science professors and (to a somewhat lesser extent) the engineering professors. These last are very close in work involvement to the entrepreneurs (the most involved of any group in the E/H career type). They are followed, in that career type, by the general managers, and then by the functional managers, the consultants, and the engineering managers. Next come the scientists in staff positions (type S/L). Not many of them are highly involved in their work, but neither are they very low in work involvement. They fall predominantly into the middle group. And finally, work involvement is lowest among occupants of engineering-based roles with low organizational positions. As a group, those in engineering staff and business staff positions tend to be rather uninvolved in their work.

Despite their generally low work involvement, however, there is a fairly sizable minority among staff engineers—in contrast with the business staff group—who are highly involved with their work. Some of these "deviate" also in their family situations: Among the 23 staff engineers with the highest work involvement (those in the top decile of the total distribution, with scores on the index $\geq$4.0), a significantly larger number than one would expect, are single, or married but childless. In this group the attributes usually associated with high work involvement are not apparent: None of the individuals is very satisfied with his job and none perceives himself as successful in his work. It is almost as if these people are involved in their work merely because they lack other involvements. In contrast, the high work involvement of the 16 staff engineers who do have children seems to represent a more basic commitment to work. Fully half of them perceive themselves as successful at work, in contrast to only 17% of all the staff engineers. All of these eight "successful" and highly work-involved engineers consider success at work to be very important, and all have very high career aspirations—they exhibit the success aspirations usually associated with high organizational roles. Though we do not know enough details about their work situations to identify facilitating factors there, we can say that all but two have a clearly technical orientation to their work. This is consistent with the finding of McKelvey and Sekaran (1977) that a technical orientation is significantly related to the job involvement of nonmanagers who were engineering majors, though not of those who were science majors. It is an important point, to which we will return in the next chapter.

The information presented so far yields the following picture of these four career types.

Type E/H This category contains 43% of the male respondents in the sample. It represents, therefore, the modal career type for MIT graduates from the 1950s at a point about 10–20 years after graduation. It consists primarily of the engineering managers and the functional managers, but includes also the entrepreneurs, the consultants, and the general managers. As a group it is governed by organizational norms, and characterized by high career aspirations and a great premium on success at work.
Type E/L This group contains 28% of the male respondents and consists primarily of staff engineers and the small group in business staff positions. It shares the value norms of the E/H group, but manifests less concern with success and careers. It is also the group with the lowest work involvement in the sample.
Type SP/H This career type comprises only 11% of the total male sample. It consists primarily of the professors and also includes the 20 science managers. These alumni share with their counterparts in engineering a high valuation of success and career, but the value norms associated with these positions are much more professional and much less organizational. This difference is reflected in the particularly high work involvement in this group.
Type SP/L This career type, which consists of the 4% of the male respondents now working as staff scientists, combines the professional values associated with the SP pattern with the lesser concern with success and career characteristic of low organizational positions.

Table 4.5 summarizes the differences in work involvement among the four career types, and also shows data on perceived success and job satisfaction,[5] in terms of the differences between the percentage of any group in the high and in the low categories. It shows clearly that occupants of engineering-based careers with low organizational positions (type E/L) have the most negative reactions to their work, with those in SP/L careers not far behind. In contrast, those in occupational roles representing high organizational positions tend to have a positive reaction to their work. Whether in the engineering or the scientific-professional pattern, their job satisfaction is about equally positive. But the success aspirations associated with these roles express themselves differently in engineering-based than in scientific-professional careers. In the former this orientation is expressed in organiza-

Table 4.5 Reactions to Work by Occupants of Four Career Types

Organizational Position	Career Pattern: Engineering-Based (E)		Scientific-Professional (SP)	
High (H)	**Type E/H (N=581)**[a]		**Type SP/H (N=151)**[a]	
	Perceived success	+18	Perceived success	+3
	Work involvement	+6	Work involvement	+27
	Job satisfaction	+12	Job satisfaction	+10
Low (L)	**Type E/L (N=376)**[a]		**Type SP/L (N=60)**[a]	
	Perceived success	−30	Perceived success	−12
	Work involvement	−17	Work involvement	0
	Job satisfaction	−31	Job satisfaction	−7

a. Reduced where necessary by number not answering all the relevant questions.

Note: Values in the table represent *D* (excess of percent positive responses over percent negative responses in distributions that have been trichotomized). The middle categories, which are not represented in the table, remain relatively constant. A two-way analysis of variance was performed with career pattern and organizational position as the independent variables, and tests were made contrasting high and low organizational position within each career pattern. For the index of work involvement both main effects are highly significant ($P<0.001$), as is each contrast, though the difference is greater in the E ($P<0.001$) than in the SP pattern ($P=0.02$). For success, only the main effect of organizational position is significant ($P<0.001$), but there is a significant two-way interaction ($P=0.004$) between career pattern and organizational position. The contrasts show that differences in organizational position are highly significant for the E pattern ($P<0.001$) but not for the SP pattern. Job satisfaction, like success, shows significant differences only for organizational position ($P<0.001$), with an almost significant two-way interaction between organizational position and career pattern ($P=0.066$). Differences between organizational position are again highly significant in the E pattern ($P<0.001$); in contrast, this difference is not quite significant ($P=0.088$) in the SP pattern. In three separate one-way analyses of variance with career type as the independent variable and with work involvement, success, and job satisfaction as the dependent variables, all differences are highly significant, with F ratios over 20 (3 degrees of freedom between groups) and with P values below 0.001.

tional terms, and is reflected in the high degree of perceived success in this group, whereas in the latter professional values are stronger and the content of the work (expressed by work involvement) is therefore more strongly emphasized.

Summary

Alumni in managerial positions and those who are consultants and entrepreneurs (type E/H) are generally positively oriented to work. Their concern with success is well matched to the requirements of their occupations and to the organizational values associated with these positions. They have the highest perceived success of any of the four types, are satisfied with their jobs, and are highly involved with their work. They contrast sharply with the alumni in engineering and business staff positions (type E/L), who represent the most problematic career type. The work roles of the alumni in this type do not seem to meet their basic needs; hence their involvement is on the whole low. They are not very satisfied with their jobs, nor do they perceive themselves as very successful. Their positions do not reflect the organizational values of advancement associated with other engineering-based careers, and their concerns lie not with success and career but more with non-work-related areas of their lives. This difference in reactions to work within the engineering career pattern is particularly striking because these two types share common educational and early-career experiences, as well as a common set of values.[6]

Within the scientific-professional career pattern, distinctions also exist. Alumni in faculty positions and science managers (type SP/H) are positively oriented to work, but in line with their greater professionalization the emphasis is more on involvement with the work itself than on organizational success. Their reactions contrast less with their counterparts in low organizational positions (type SP/L) than is true for the equivalent contrast in the engineering pattern. Staff scientists are basically in the middle range on all these indicators of their reactions to their work.

Thus, there are very general differences in reactions to work among these four career types. But there are also differences among the individuals within each of these types. The characteristic profiles identified in this chapter represent modal tendencies, and in each type of career there are exceptions, people whose reactions are different from those of most of their colleagues. Not all engineers in staff positions are unhappy. Indeed, as indicated, some are very much involved with their work. It is not position alone, therefore, that determines a person's reactions to his work, but also how the requirements of an occupational role fit with his particular values, aspirations, and skills.

Notes to Chapter 4

1. In terms of complexity and structure, this index resides somewhere between the extremes of single-item indicators and qualitative typologies. It therefore combines the analytic advan-

tage of a structured instrument and a large sample with the attempt (usually associated with more qualitative techniques) to reflect the situation from the point of view of the respondents. A sophisticated example of the use of a single item is given by Vroom (1962). His concept of "ego-involvement" is described in terms not too different from those we use, but his measure is based on a single item with 4 levels of response: "If a problem comes up in your work and it isn't all settled by the time you go home, how likely is it that you will find yourself thinking about it after work?" For purposes of comparison and change over time, of course, such single-item indicators are invaluable, as is obvious by the frequent reference made to the early use of this approach by Morse and Weiss (1955). Kahn (1972) gives an excellent review of such single-indicator measures. At the other extreme we can mention Sofer (1970), whose classification of some 80 executives and technical specialists in two industrial organizations is based on responses to four open-ended questions ("Can you give me an idea of your other main activities and interests outside your job?" "What part do you feel your work plays in this overall pattern?" "If you can imagine for a moment that financially you didn't have to work, what do you think you would do?" and "If you did give up your work here, is there anything you would miss?"). For a full review of the literature on job involvement and related concepts see Rabinowitz and Hall 1977.

2. It must be stressed that this is a probabilistic classification. It is not based on an individual analysis of each alumnus's organizational position, but rather on the assessment that positions in a particular occupational role are likely to be high or low. This distinction, therefore, applies to occupational roles and not to individuals. The assumption is that most of the individuals in an occupational role classified as high will have high organizational positions, and that most of those in occupational roles classified as low will be in low organizational positions. But the probability of such an association, though high, will be less than 1.

3. When one looks at the overall figures for each career type, one sees much greater difference in values between the successful occupants of engineering-based and scientific-professional careers than between those in high and low positions. The emphasis on organizational values of advancement and earnings is almost the same in E/L careers as it is in E/H positions, and SP/L occupations emphasize professional rather than organizational values to the same extent that the faculty positions do. The only exception on the engineering-based side is the staff engineers, who are less inclined to support the organizational value syndrome than are the other engineering groups. But even they are more likely to adhere to these values than are the staff scientists, their SP equivalents. And, on the scientific-professional side,

only the science managers among the SP/H careers deviate somewhat from this pattern; they are more organizationally and less professionally inclined than the other groups in the SP pattern. But even they are less organizationally inclined than are most of those in engineering-based careers. (These conclusions are based on averaging the percentages given in table 3.3 over items and, where relevant, over occupational groups. See also Bailyn 1977b, table 2.)

4. In high organizational positions 36% of those in the E pattern and 34% in the SP pattern have high success aspirations. (The equivalent figures for low success aspirations are 34% and 35%.) In low organizational positions, in contrast, only 15% of E-pattern alumni and 17% of SP respondents have high success aspirations, compared to 61% and 58%, respectively, with low success aspirations.

5. The measure of perceived success is described in appendix C, and the definition of "high" success is the same as used there. Respondents in the "low" success category are those whose ratings of their current success are ≤ 3 on a 1–5 scale: 34% fall into this low category, which matches exactly the 34% in the high category. Job satisfaction is gauged by a direct overall question with a 1–5 response scale. All those (27% of the sample) who rate themselves as very satisfied with their jobs ("5") are in the high category; those whose ratings are ≤ 3 (29% of the total) are in the low category.

6. However, as already indicated, these values are not as strong in the engineering staff group as in the other engineering-based occupations.

5 Individual Orientations and Organizational Career Paths

The modal profiles identified in chapter 4 begin to pinpoint certain types of careers that seem to "work" at mid-career and others that are likely to present difficulty to the people who pursue them. In particular, they show that high organizational positions are experienced more positively than are low organizational positions, in both the engineering-based and the scientific-professional career patterns. Of course, under current organizational procedures such "high" positions pay more. But, as one 1959 graduate who quit his job at a time when he had just been "given more money and stock options" said, "Money alone does not make the job. I am not indifferent to money, but a job should be fun if one is to spend his life at it."

And even more than fun is at stake. Graduates committed to technical careers from an early age and concerned with the challenge and creativity of their work are not likely to be satisfied for long in positions in which their autonomous contributions are not encouraged and recognized. They are likely to complain about such jobs, as does a 1959 graduate from the department of chemical engineering, now a general manager, who says that "an engineer or scientist functions as, is treated as, and is hired as a tool (or a mechanic) to accomplish a specific objective under the responsibility of management."

It is the invidious distinction between those in management and those not that seems to be the crux of the difficulty. Such a simple bifurcation of organizational career paths is incommensurate with the wide variety of expectations and capabilities that technically trained people bring to their jobs at mid-career. This chapter pursues this point by considering whether there are systematic congruences or incongruences between individual orientations and the way organizational roles are currently defined. In particular, it treats the question of what kinds of people do or do not fit each of the four types of careers previously identified. The findings suggest a number of changes in the assignment of organizational roles to technical personnel that could benefit both employees and employers.

Individual Orientations to Work

Gordon (1977) found that concern with the management of people and concern with the management of technical problems represent two different orientations which are differentially satisfied by different kinds of jobs. And results already presented show a technical orientation for alumni in low-level engineering-based jobs who, contrary to the modal reaction in that career type, are involved with their work. These findings indicate that the distinction between a technical and a more human orientation is relevant to the fit of individuals to their jobs.

Not every MIT graduate at mid-career, however, has a well-defined orientation to work. In fact, some of these respondents have withdrawn from involvement with work altogether and are instead oriented to family and other non-work-related areas of their lives (Bailyn 1977a). Such employees are not necessarily ineffective, as long as they are assigned to tasks commensurate with their needs. Hence, it is important to distinguish this relatively uninvolved group from those whose orientations run along specific work lines. The subsequent analysis, therefore, compares three groups: individuals who are relatively uninvolved in their work; those whose work involvement is oriented toward people; and those whose work involvement is oriented to technical problems.

The details of the derivation of the measures of these orientations are given in appendix E. Generally, the low end of the index of work involvement is used to identify non-work-oriented alumni. Orientation toward people ("human orientation") is gauged by each respondent's reply to a direct question about the importance he places on working with people. Indication of a technical orientation is indirect; it is based on items in the questionnaire shown by previous analysis to be related to such an orientation (Bailyn 1977b).

Initially, every respondent who answered the necessary items in the questionnaire was classified separately according to whether each orientation was high, medium, or low. With this approach it is obvious that some people can be equally high on more than one of these orientations, and others may not be high on any. Though these are interesting possibilities in their own right, and with better measures would be worthy of further analysis, the emphasis in this section is on contrasting the orientations.

Table 5.1 Distribution of Unambiguously Oriented Individuals

Primary Orientation	N	% of Total[a]
Technical	260	21%
Human	219	17%
Non-Work	248	20%

a. The total number on which these percentages are based is 1,264 and excludes those who were unascertainable on any of the orientations.

Table 5.2 Orientations in Four Career Types

	Career Pattern	
Organizational Position	Engineering-Based (E)	Scientific-Professional (SP)
High (H)	**Type E/H (N=302)**[a] 22% technical 44% human 34% non-work[b] (45% ambiguous)	**Type SP/H (N=96)**[a] 77% technical 13% human 10% non-work[b] (31% ambiguous)
Low (L)	**Type E/L (N=196)**[a] 27% technical 19% human 54% non-work[b] (44% ambiguous)	**Type SP/L (N=27)**[a] 81% technical 4% human 15% non-work[b] (52% ambiguous)

a. N values represent those who are unambiguously oriented.
b. These percentages do not match those with low work involvement in table 4.4 because they are based only on those alumni whose orientations are not ambiguous and because the cutting point was adjusted to match the other orientations (see appendix E).

For this reason, the analysis is limited to those who are classified as high on one of these orientations and not equally as high on either of the others (see appendix E for details). This advantage in clarity, however, means that the sample is considerably reduced. Of the 1,264 people who could be classified on all three orientations, only 727 (58%) are unambiguously oriented in this sense.[1] They are distributed as shown in table 5.1.

If one concentrates on those respondents who clearly have a primary orientation, each career type appears to have a distinctive orientation associated with it (see table 5.2). Alumni in scientific-professional careers are predominantly technically oriented, those in engineering-based careers associated with high organizational positions tend to be oriented toward people; and those in engineering-based careers with low organizational positions are primarily non-work-oriented. These characteristic orientations are consistent with and expand on the modal profiles outlined in the preceding chapter. In this chapter, however, the goal is to see how orientations, both typical and "deviant," fit with occupational requirements.[2]

The Fit Between Individual Orientations and Career Type

In pursuing the question of fit, one must look at both employers' and employees' viewpoints. On the organizational side, one would like to know how effective employees with various orientations are in each type of career. Though the data do not include this information explicitly, it is possible to use salary as an indirect indicator of performance if one makes it comparable across occupational and age groups.[3] Total professional income, if normalized by occupational role and class of graduation, allows one to identify people whose compensation is above or below the average of the group of respondents closest in age and in the most similar kinds of jobs. The assumption is that employees whose salaries are above average by this criterion are the more effective employees.

Table 5.3 shows that in engineering-based jobs with high organizational positions, a human orientation not only predominates but is also most highly rewarded by the organization, though a technical orientation is not far behind. Non-work-oriented employees are considered least effective in this group, at least as judged by normal-

Table 5.3 Organizational Evaluation: Percentage with Above-Average Incomes in Each Career Type and Orientation

Organizational Position	Career Pattern: Engineering-Based (E)		Career Pattern: Scientific-Professional (SP)	
High (H)	**Type E/H**		**Type SP/H**	
	Technical:	37% of 59	Technical:	46% of 70
	Human:	41% of 123	Human:	40% of 10
	Non-Work:	26% of 95	Non-Work:	20% of 10
Low (L)	**Type E/L**		**Type SP/L**	
	Technical:	63% of 49	Technical:	45% of 20
	Human:	29% of 35	Human:	[N=1]
	Non-Work:	43% of 94	Non-Work:	[N=4]

ized income. A similar pattern exists for the scientific-professional roles. There the technical orientation, which is the dominant one, is most rewarded. Among SP alumni in high organizational positions, the human orientation is close behind the technical one and non-work orientations are penalized just as in engineering. In other words, in all "high" organizational roles a definite work orientation is seen by employers to be more effective than a non-work orientation. The situation in the "low" engineering-based occupations is more complex. As might be expected, high relative incomes clearly go with a technical orientation. But a non-work orientation here is not penalized as much as a human orientation. In this group it is the employees oriented to people who are least likely to have above-average incomes. Thus, organizational evaluation—at least as reflected by relative income—seems to depend not only on orientation as such but also on the degree of congruence between orientation and career type.

In order to see how orientations fit with career types from the individual's point of view, we used the question on job satisfaction to see which combinations are more or less likely to elicit responses indicating satisfaction. Table 5.4, which presents this information, shows that about a quarter to a third of the respondents in scientific-professional careers are very satisfied with their jobs no matter which orientation they have. However, in the engineering-based careers job satisfaction is highest among the technically oriented. Almost a quarter of the technically

Table 5.4 Individual Satisfaction: Percentage "Very Satisfied" with Their Jobs in Each Career Type and Orientation

	Career Pattern			
Organizational Position	Engineering-Based (E)		Scientific-Professional (SP)	
High (H)	**Type E/H**		**Type SP/H**	
	Technical:	53% of 64	Technical:	31% of 74
	Human:	31% of 133	Human:	25% of 12
	Non-Work:	19% of 104	Non-Work:	30% of 10
Low (L)	**Type E/L**		**Type SP/L**	
	Technical:	23% of 53	Technical:	36% of 22
	Human:	10% of 38	Human:	[N=1]
	Non-Work:	9% of 104	Non-Work:	[N=4]

oriented alumni in "low" engineering-based careers are very satisfied with their jobs. This is almost as high a level of job satisfaction as is exhibited by the alumni in "high" engineering-based roles who are oriented to people. Even there it is the technically oriented who are most likely to be very satisfied with their jobs.[4] We will return to this point in the concluding section of this chapter.

Finally, in the analysis of fit between orientation and occupational requirements it is important also to look at individual satisfaction when organizational evaluation, as indicated by relative income, is held constant. (This discussion must be limited to the engineering-based careers, because the number of non-technically-oriented alumni in the SP pattern is too small for such an analysis.) Table 5.5 presents the data. In both E/H groups (with above average or below average incomes) technically oriented alumni are more likely to be very satisfied with their jobs than are alumni oriented toward people. However, in the "low" engineering roles the picture is different: Although the technically oriented respondents in the E/L career type are on the whole evaluated positively by their employers (as shown by their above-average salaries), these relatively well-paid, technically oriented employees in E/L positions are not particularly satisfied with their jobs. On the contrary, the satisfaction level of this subgroup is no higher than that of alumni in this career type with other orientations. In fact, only among technically oriented employees who have below-average incomes does one find a sizable group very satisfied with their jobs. These people, technically oriented and of average capability, are not insignificant to their employers, as they probably do most of the routine technical tasks, and their relatively high satisfaction presumably contributes to their effectiveness. What is problematic for organizational policy, however, is the fact that the best technically oriented people in the E/L group (at least, those most highly rewarded) show so little satisfaction with their jobs. This seems to indicate that engineering-based staff positions are not set up to meet the needs of the most competent technically oriented employees. Despite the fact that most of the technically oriented people in these jobs are evaluated positively by their organizations, they are not satisfied with their positions. This finding should challenge organizations to think more creatively about their

Table 5.5 Percentage "Very Satisfied" with Their Jobs in Career Types E/H and E/L According to Relative Income and Orientation

	Relative Income			
	Above Average		Below Average	
Type E/H	Technical:	62% of 21	Technical:	54% of 26
	Human:	39% of 51	Human:	24% of 50
	Non-Work:	28% of 25	Non-Work:	16% of 51
Type E/L	Technical:	13% of 31	Technical:	31% of 13
	Human:	10% of 10	Human:	11% of 19
	Non-Work:	10% of 40	Non-Work:	9% of 44

technical staff roles and how they can be made more attractive to high-potential people whose orientation, at mid-career, remains technical.

Thus, there is evidence in the engineering-based careers—where high positions tend to be managerial—of some lack of congruence between individual orientations and organizational career paths. In contrast, in the scientific-professional career pattern—where high positions are represented primarily by university professorships—there seems to be a relatively good fit between individual orientations and occupational requirements.

Four Career Types: A Summary Profile

The modal career type for MIT alumni of the 1950s is type E/H, which is engineering-based (in terms of undergraduate major and early career experiences) and is associated with high organizational positions, primarily as technical or other functional managers but also as entrepreneurs, consultants, and general managers. It represents the traditionally "successful" organizational career path for engineers. And, not surprisingly, more of the alumni who have traversed it perceive themselves as successful than in any other career type. Perhaps the only surprising finding about this group is that although its modal orientation is toward people, this characteristic stance is not the one associated with the highest job satisfaction. These high organizational positions, defined as involving primarily the supervision and development of people, are in fact most satisfying to technically oriented alumni. This finding is consistent with other studies of engineers (e.g., Ritti 1971; Thompson and Dalton 1976) and suggests that high positions are satisfactory to the technically oriented in this group because they give them the power to be autonomous in their technical functioning. It seems to be the *autonomy* of high organizational positions that is important for this group, not the prestige or tasks associated with the management of people. It is at least curious, and perhaps a serious flaw in organizations, that this kind of autonomy is often not available to technically oriented employees without the accompanying duties of a manager.

The point is confirmed from the negative side by the second most frequent career type in the sample, type E/L —engineering-based careers whose occupants are still, at mid-career, in staff positions. This group has the most negative reactions to work of any: the lowest perception of their own success, the lowest work involvement, and the lowest job satisfaction. Indeed, their modal orientation is a non-work orientation. Those in this group who are oriented to people seem to be particularly mismatched with their roles. Only among those who are technically oriented does there seem to be any fit at all. And here one is confronted with the anomaly that only among technically oriented people whose relative incomes are below average does job satisfaction approach that of any other career type. This finding corroborates the dysfunctions associated with the way technical roles are currently organized. High-potential technically oriented employees, it seems, either have to switch to managerial roles or must remain in roles that do not provide them with enough autonomy for satisfactory expression of their technical competence.

In the scientific-professional career pattern, those in roles associated with high organizational positions (type SP/H)—mainly alumni who went on to get their doctorates and are now science and engineering professors or science managers—seem to fit well the requirements of their jobs. They have the highest work involvement of any career type, and both their perceived success and job satisfaction are high. Their orientation is primarily technical (it is of interest, too, that they have the lowest percentage with ambiguous orientations), but they seem to be equally satisfied with their jobs no matter what orientation they have. There is also only a slight difference in relative income between those with a technical and those with a human orientation; non-work-oriented alumni in this career type, though, are less likely to be in the above-average income group.

Finally, scientific-professional alumni in roles associated with low organizational positions—those who are staff scientists—show a high amount of ambiguity of orientation, but those who can be classified are almost exclusively technically oriented, and these are as satisfied with their jobs as are their counterparts in high organizational positions. They are not, however, as work-involved; nor do they perceive themselves as successful. But they are considerably more positive on all these indications of their reactions to work than are alumni with equivalent organizational positions in engineering-based careers.

It is clear from the profiles of these four career types that a satisfactory work situation at mid-career results from a complicated interplay of conditions involving the fit between individual orientations and the structure of jobs within organizations. In the concluding chapter the implications of these findings are spelled out in great detail. It suggests a much greater differentiation of organizational roles in the deployment of the technical work force. In particular, as will be seen, it suggests that responsibility and recognition should not be limited to managerial positions but should be allocated also to a variety of technical roles.

Notes to Chapter 5

1. Another 327 (26%) are not high on any of the orientations, and 210 (17%) are equally high on more than one. In the latter group it is primarily a human orientation that combines with either a technical or a non-work orientation. Only 27 respondents combine a technical and a non-work orientation, and only 20 are equally high on all three.

2. The number of SP/L respondents who are unambiguously oriented toward people (1) or who are non-work-oriented (4) is so small that this career type cannot be included in the analysis.

3. The normalization was done with total professional income over 36 groups: 12 occupational categories, each further subdivided into 3 graduating classes. This procedure gave each respondent a score indicating the number of standard deviations he falls above or below the mean of his occupational/class group. Of the 1,094 respondents available for this analysis

(which excludes those who did not fit into the 12 occupational categories used—the 11 described in chapter 2 plus the architects—as well as those who did not give salary information), 15% fall at around the mean, 42% have incomes above the average of their occupational/age groups, and 43% are below average.

4. This is true for every engineering-based occupational role except the general managers. Only in this group are the alumni with a human orientation (N=21, of whom 48% are very satisfied with their jobs) more satisfied than those whose orientation is technical (N=3, of whom 1 is very satisfied).

6 The Women Graduates

During the 1950s less than 1% of each MIT graduating class was female. The Institute put no official constraints on the admission of women, but there were no inducements either—there were no women's dormitories, for example. The numbers were small primarily because few women applied.[1] Not only would a woman have had to show exceptional ability in mathematics and science to be considered eligible for MIT; she also would have had to be willing to put herself into a very male environment. Some scientifically inclined women probably preferred the atmosphere of a women's college and expected it to be more conducive to their intellectual development in a "masculine" field;[2] others may simply have felt uneasy about being so heavily outnumbered (Laws 1975; Kanter 1977). Further, MIT at that time was much more specialized than it now is, requiring a stronger commitment to a specific set of fields at a relatively early age.

And yet, for the women students who came there were obvious attractions. One woman puts herself in the category of "people who should never have gone to MIT in the first place, but who, after ten years, have finally 'found' themselves in another field," and strongly recommends pre-entry counseling and testing of women applicants "to preclude the re-occurrence of situations like my own." Yet she recalls that "the notion of being a woman student at MIT is a highly seductive one—it's quite possible that no amount of counseling could have kept me away." And another alumna, who remembers her childhood as "full of bitter experiences of being denied odd jobs because I was a 'little girl,' " notes that "MIT probably saved me. I might have been the first female assassin because I was very angry."

Whatever their motivation in coming to MIT, it is clear that these pioneering alumnae are exceptionally able women, and were well educated for careers in their chosen fields. They graduated, however, into a world of the "feminine mystique," where behavior was based more on traditional sex-linked expectations than on individual capabilities and inclinations. But though their task was different from that faced by women today, some of their experiences may well be instructive for younger women (and men), who are also faced with issues of choice and accommodation in their lives.

Career Paths

Of the 22 female graduates in the classes of 1951, 1955, and 1959, 15 responded to the questionnaire. Of these, 5 graduated from the School of Science (a higher proportion than among the men), 7 from the School of Engineering (a proportion somewhat lower than among the men), 2 from the School of Architecture and Planning, and 1 from the School of Humanities and Social Science. Two have doctorates and 5 have master's degrees.

During the first 2 years after getting their BSs, 13 of these 15 respondents had what might be considered standard post-MIT experiences:[3] Two were full-time graduate students throughout the 2 years, 7 were employed by private profit-making companies (6 as engineers, 1 as a scientist), 2 were in nonprofit research institutes (1 doing research and 1 working as an engineer), and 2 were with private architectural firms (1 only as a draftsman).

The first post-B.S. experiences of these alumnae, then, were very similar to those of MIT alumni in general. But when one looks ahead a few years, the various patterns of women's careers begin to emerge. Five years after graduation, only 6 of these 15 alumnae were pursuing careers commensurate with the education they had received. Four were not working at all, and 5 were working but had restricted their work in various ways in order to accommodate their family roles.

These career styles did not change very much after this 5-year point. At the time of the survey, some 11–19 years after graduation, only one more alumna had joined the ranks of those with full participation in a career: the one, already mentioned, who found her field a decade after graduation and ended 10 years of accommodative jobs by entering a social science Ph.D. program (and, incidentally, getting married for the second time). Only two other changes in the level of career participation occurred after 5 years beyond graduation. One woman, who was not employed for 10 years while she raised three children, subsequently re-entered school to pursue a degree in teaching and at the time of the survey was teaching high school mathematics, thereby changing her category from "no work" to that of an "accommodative job." A reverse shift was made by the alumna who worked until her first child was born (8 years after graduation and 6 years after her marriage) and then left the job market altogether.

Thus, at the time of the survey 7 of the 15 female respondents were participating fully in careers generally typical of MIT graduates (including the woman referred to above who was in a full-time doctoral program and one who, though currently unemployed because her family had recently moved to a different part of the country, was actively seeking employment). Four of the respondents had no careers at all—only one of these (an army wife who had spent a great deal of time overseas) was truly satisfied with this state of affairs; the other three showed some longing for more participation in the future, and two of them expected in fact to have it when their children were grown. The other 4 respondents pursued what we have called an accommodative career. Various strategies are available for such accommodative ways of coping; some work part-time in their fields, others restrict the scope of their possibilities (as the architectural designer who never got her certification as a registered architect) or work in a "lesser" job than the sort for which they were trained (as a crystallographer working as a physics research editor whose task is to "verify facts in physical sciences").

Table 6.1 summarizes this information and shows the relation of career style to family status. It is obvious from this table that the degree of career participation of these women is directly related to their family status. All of the single women and the married ones with no more than one child participate fully in careers; none of those with three or more children do. Those with two children—the modal pattern in the sample—are represented in each category, though skewed away from full participation. In a previous era marriage itself was the major barrier to a

Table 6.1 Career Style and Family Status Among Alumnae

	Career Style		
Family Status	Full	Accommodative	None
Single	3		
Married, no children	1		
Married, 1 child	2		
Married, 2 children	1	2	2
Married, $\geq$ 3 children		2	2

woman's full participation in a career, but this is obviously not the case for this group. Now children are the barrier, but their number is apparently more important than their presence. One child can evidently be assimilated relatively easily into a full career; two are more difficult; three or more are almost impossible.

The timing of children seems to be unrelated to future career style. Though all those alumnae who were to marry did so within 2½ years of graduation, not all had children right away. Delay in having children, however, is not associated with greater career participation. Two of the respondents with full careers gave birth within a year after marriage, the respondent whose children were delayed longest (13 years) had only accommodative jobs even during her childless years, and of the two respondents who waited 5 years before having a first child one ended up with a full career and the other with none.

These career styles are also related to these women's educational and early-career experiences. In a way the differences were already nascent in college. All of those whose grade-point averages were 4.0 or more (on a scale to 5) ended up with full career participation; not one of those who ended up without any participation in work had an average over 3.5. Graduate work is also related to eventual career style. None of those currently not employed attained a higher degree (though they may have taken some graduate courses); both of those with work on the doctorate level became full career participants. And though the initial postgraduation experiences were not greatly different from those of the men, the fate of these alumnae's first jobs foreshadowed their eventual career styles.

All of the women who stayed on their first jobs for more than 5 years ended up with full careers. Only one single and one married woman with full career participation left their first jobs earlier (whereas none of those in the other groups stayed more than 2½ years in their first jobs). Both of these left primarily because the job did not offer challenging work or a personal sense of accomplishment; as secondary reasons the single alumna blamed the lack of freedom to adopt her own approach to the job and the married one deplored her lack of opportunity to advance. The two alumnae who ended up with no work participation but who were not married during the first year after graduation show an interesting contrast. One also left her first job primarily because it did not give her enough challenging work to do (and secondarily because it did not allow her to contribute to society). But she left this initial position after a little over a year and took a clearly more accommodative job (she went from being an engineer in a private company to being an administrative assistant to the chairman of an engineering department in a university). And this job she left a year later in order to get married. The second alumna left her job as an engineer in a private company because she did not like its location, and moved to another engineering job for 2 years, during which time she married. The other two alumnae who ended up with no work participation were already married when they started their first postgraduation jobs, and both left them when they got pregnant. All of those who ended up adopting an accommodating strategy left their first jobs to follow their husbands to another location.

In other words, the career paths of these able women, though initially established in accordance with their talents and interests, were soon modified by their work and family experiences. Indeed, the eventual career style of the alumnae appears to have been embedded in the experiences of their first few years after graduation. But since these women confronted upon graduation expectations and opportunities very different from those faced by today's women graduates, the implications of this finding are not obvious. We will return to this point in the concluding section of this chapter.

Midlife Career Styles

In order to understand the careers of these alumnae and their reactions to work at mid-life, it is necessary to look separately at those with full careers (married and single), those with accommodative jobs, and those not employed.

Full Careers Table 6.2 shows the career types and the reactions to work (according to the same scheme used for the male alumni) of all the employed alumnae. The occupational categories of the fully participative alumnae are not too different from those of the men, though in every case except architecture the incomes of these women are well below the average for their occupational/class group.

Table 6.2 Career Style and Reactions to Work of Employed Alumnae

	Occupational Category	Career Type	Normalized Income
Full Careers	Staff engineer	E/L	−1.0
Single	Staff engineer	E/L	−1.0
	Architect	–	+1.7
Married	Functional manager	E/H	−1.2
	Engineering professor	SP/H	−0.8
	Science professor	SP/H	−0.5
	Social scientist	–	NA[b]
Accommodative Careers (all married)	Research editor		
	Part-time architectural designer		
	High school teacher		
	Part-time lecturer		

a. On ascending scale of 1–5.
b. No answer.

A comparison between the fully participating married and single alumnae yields some interesting differences. The single women, who presumably share with the male alumni the need to work even if their motivation is not very high, seem less work-oriented than the married ones, for whom career style is more nearly a matter of choice: The married women are higher on both technical and human orientation. And though the level of currently perceived success is about the same, the single alumnae feel they have reached their peak; none expects her success at work to go higher. The married women, in contrast, all expect to be more successful at the height of their careers than they now are. They seem still to be in an upward cycle in their careers. And this is true even though the first three married women in the table match almost exactly the ages of the single alumnae. The women in both these groups are fairly professional. All belong to professional societies; all but one (the architect) have published. But only two (both married) have read a paper at a professional meeting.

This small group of married women (it must be remembered that only one has two children; the others have one or none) seems to have been able to reach, at mid-life, a

Work Involvement	Technical Orientation	Human Orientation	Job Satisfaction[a]	Perceived Success[a]	
				present	future
medium	low	low	2 (low)	3 (low)	3
high	low	low	4	4	4
medium	high	low	4	4	4
medium	very high	very high	4	4 (high)	5
high	very high	very high	5 (high)	4 (high)	5
NA	NA	medium	NA	3 (low)	5
medium	very high	high	NA	2 (low)	4
medium	very high	medium	4	4	4
medium	NA	NA	5 (high)	3 (low)	3
medium	high	high	4	3 (low)	4
medium	high	very high	5 (high)	4 (high)	5

satisfactory integration of work and family. None moved into her present position in a linear fashion: All had "slow" periods in their careers (see Bailyn 1979), and all have modified their initial career lines at least to some extent. Though they may have paid the price for these "deviations" in lower-than-average salaries, their single counterparts, whose careers have been more "orderly" (Wilensky 1961), have not done much better in this respect and have arrived at mid-life less optimistic about the future of their careers.

Accommodative Careers Two of the four women in accommodative careers seem to have chosen to manage their lives in this way; the other two seem to have been forced into this career style by circumstance, against their hopes and expectations. The former two are the more satisfied. Both have published, though neither now belongs to a professional society. The latter two, in contrast, both belong to professional societies, though only one has published. These accommodative alumnae (all of whom have either two or three children) share with their married classmates with full careers an orientation toward both technical and human concerns. And though they are not as optimistic about their future careers as this group, they are more optimistic than the full-career single alumnae. But their job values are different from both fully participating groups—they are more indicative of their basic accommodating strategy: "reasonable workload," "sufficient time for family and personal life," and "location" are all particularly important to them.

In many ways the careers of these alumnae are not comparable to those of the men in this study. But this situation may change for later graduates. Today it is not unheard of for men, particularly if they are part of dual-career families, to want their careers to be accommodative to family needs (Bailyn 1978). The somewhat greater job satisfaction and career optimism of these accommodative alumnae as compared with the single women whose careers followed more traditional "masculine" lines is, therefore, of some interest.

No Careers The group not currently working is pretty much as expected. Their families are their primary sources of satisfaction in life; they display no career orientations and have no professional connections. Their only two positive "job" values are "feeling of accomplishment" and "time for personal life and family." When asked to think back to their undergraduate years these alumnae mainly complain about the lack of practical education they had at MIT. An "electrical engineer" with three children, not working currently and with a not-very-good job record after her graduation, complains that MIT was "too theoretical . . . no knowledge of how to begin to solve a problem or design a circuit"; a math major with a checkered work and graduate career complains of the "lack of balance between theoretical and applied math" and says that "the math student who could not immediately attend graduate school was not properly prepared to work in industry." But none of these women feels that her education was wasted. One wrote:

> I think my education was worth the time and expense and have not regretted it for a minute. It is "insurance" against any time I may have to work again, it has aided me in raising the children, . . . because of my scientific education I have been able to communicate more fully with my husband who . . . specializes in research [and] computer work . . . .

Conclusion: Future Trends?

It is obviously difficult to say anything definitive on the basis of 15 cases, particularly when they fall into four different categories. But these women are pioneers. And though they established their life styles in a world less malleable than the one we know today, their experiences are at least suggestive.

The findings suggest that full careers along traditional lines are only a realistic option for women with no children, or at most one. It is a real question whether today's availability of child-care facilities and the greater tendency for fathers to participate in child care will ease this constraint on women's careers or whether the latter will make it more difficult for men with two or more children to traverse traditional career paths to the top.

The data also show that length of time in the first job is an important element. The importance of the early career years has often been emphasized for traditional career paths, and indeed those women in the sample who had to leave their first jobs in order to relocate with their husbands found themselves unable to establish full careers. But rather than accepting this as a fact of organizational life and trying to fit people into these traditional career lines, it may be well for employers to consider alternate paths for successful organizational careers. In this respect, the differences between the fully participating married women and the single alumnae are particularly instructive. Organizations value involved and optimistic older employees. If they realize that the probability of these responses is greater after less linear or orderly careers than they seem to insist on at the beginning, the association between career success and early-career events may lessen.

Accommodation, though sometimes a response to adverse career experiences, may also represent a positive choice for some—men as well as women (Bailyn 1977a). And part-time work or other modes of career restriction may make sense in an economy whose growth is slowing down. The accommodative career style may therefore become more prevalent for both sexes, particularly when two modified careers can provide the same standard of living as one full career.

Combined with the increasing number of dual-career families, these findings imply that both men and women need more choice in career paths, more options in their work lives. Thus, organizations that further the belief that the orderly, linear career is the only way to succeed (whether by formal policies or by the assumptions underlying personnel procedures) may not be making the most effective use of their work force.

Notes to Chapter 6

1. There were 125 female applicants for the class of 1959 and 3,655 male applicants. Of the 114 decisions made on women, 32 women (28%) were admitted, of whom 21 registered (a yield of 66%). More of the actions on applications by men were favorable (48%, 1,654 out of 3,452), but the yield was somewhat less: 900 (54%) registered. A more striking difference is obvious when the graduation figures are considered. Only about one-third as many women graduated as had registered as freshmen, in contrast with a ratio of graduates to matriculants among the men of approximately 7:8.

2. David Riesman (1970, p. 53), commenting on a 1968 Smith College survey on attitudes toward coeducation, reported that "girls majoring in the sciences or mathematics were most apt to have come to Smith originally because it was not co-ed, and were most apt to prefer that Smith not go co-ed now."

3. Of the other two, each of whom was married and had a child during this early period, one had a job as a technical editor in a private research organization but 10 years later was fully immersed in a new field and the other was working part-time during this period but within a few years went back to get a graduate degree and then took a full-time job teaching at a university. It is perhaps significant that these two, who started their careers in more peripheral ways, both eventually attained full careers. This attests to the possible advantages for mid-career of slow early starts.

7 Implications and Conclusions: Toward New Organizational Roles

At this point in my life I would gladly trade some of my professional success for more success in the rest of my activities.
—systems analyst, 37 years old

This statement by one of our respondents is not atypical. It must be noted, however, that such comments are not distributed randomly throughout the sample, but are particularly characteristic of one type of career. It is not merely chronological age—the arrival at mid-life—that produces concern about the value of one's career. Rather, it is more the particular pattern of work experiences, the way in which technical jobs in organizations are structured, that makes certain careers unsatisfactory for long periods of time:

The greatest lack and disappointment in my technical career is the impossibility of achieving security and recognition without leaving it.
—development engineer, 40 years old

That the structuring of technical work has an impact on career outcomes and on people's feelings about the role of work in their total life is one of the main findings of the research reported in this book. It is a conclusion that carries with it some optimism, for although one cannot stop the process of aging, one can reexamine the organizational assumptions and procedures that define career paths, and thus at least begin the process by which counterproductive policies might be changed.

Summary of Findings

Before turning to these organizational implications let us review the results briefly. We concentrate here on the findings from the male alumni, the bulk of the sample. The data from the women graduates (chapter 6) serve as a useful counterweight to this picture, as a clue to alternative ways of approaching technical careers, and will be referred to in the concluding note.

The analysis started with a classification of current jobs, which clearly reflects the technical background of the sample. Over 50% of the respondents still hold technology-centered jobs 11–19 years after graduation. But within this set of occupations there is considerable variety—consulting, bench-level engineering, technical management, university teaching. Such technical careers can be played out in many ways and in many settings. They can involve bachelor's, master's, or doctoral degrees; they can be combined with management and yet retain their technical core; and they can be primarily scientifically based or based in a more applied engineering discipline. These technical backgrounds can also lead to managerial and business careers, and indeed are the source of some of the highest entrepreneurial and general-management jobs in our society. For example, of the 50 general managers in the sample, 12 are presidents or executive vice-presidents of corporations with at least 2,500 employees, and two of these (both in their early forties) have top-line responsibility in corporations of 10,000 employees or more.

From this initial occupational classification we then moved backward in time and found that on the basis of undergraduate major, academic performance, level of degree attained, and first job, two major career patterns emerged: an engineering-based pattern, which follows fairly directly from an undergraduate degree in engineering and leads to many kinds of occupations, some of which involve abandoning the technical core altogether; and a scientific-professional pattern, characteristically involving a doctorate and pursued in technical work, technical management, or academic activities. It seems to be the level of professionalism represented by the doctorate that most clearly distinguishes the two patterns. Thus, engineering graduates whose good academic records as undergraduates propelled them to continue their education to the doctoral level tended to become academics, and, as engineering professors, share more of the characteristics of the scientific-professional than of the engineering-based career pattern. These two patterns were also shown to be associated with distinct sets of career values. The "organizational value syndrome" identified in the engineering pattern centers on the importance of contributing to one's organization and on the desire for high earnings, advancement, and leadership. In contrast, the "professional value syndrome" characteristic of the scientific-professional career pattern centers on accomplishment and challenge, on the opportunity to improve one's knowledge and skills and to use them creatively.

This patterning is significant because it indicates that technical occupations are neither as homogeneous as some have suspected (scientists and engineers, for example, even in the same industrial context, are different in important ways and probably need to be managed differently) nor as individually distinct as others might assume.

The next step in the analysis was to turn to the respondents' reactions to their work. Here we found systematic differences among the occupational categories within each career pattern. By differentiating the occupations associated with high work involvement from those where involvement is typically low, we came to a further, independent distinction: a classification of the occupational roles according to whether the organizational positions associated with them are more likely to be high or low. That this distinction is meaningful was shown by the fact that respondents in roles classified as low in organizational position had considerably lower success aspirations than did those in roles classified as high. Since there was no difference between the two career patterns in such success aspirations, and no systematic difference in value syndromes between high and low organizational positions, we had two independent ways of classifying occupational roles and could put them together to form four career types. The analysis of these career types has shown that each presents a different modal profile.

The most problematic career type is type E/L, the engineering-based careers with low organizational positions. Here one finds the lowest work involvement, the lowest perceived success, and the lowest job satisfaction. In contrast, the two high-level types, E/H and SP/H, are more positive in all of these reactions, but, in line with the associated value syndromes, the E/H group is highest in perceived success whereas the SP/H group is highest in work involvement. Further, the scientific-professional groups are primarily oriented toward the technical aspects of work, whereas among the engineering groups orientation depends on organizational position—if it is high the primary orientation is toward the human aspects of work; otherwise the orientation is not toward work at all. Gearing an education to the production of "engineers" implies at least the chance for a successful lifetime occupation. This research raises the possibility that such a lifelong occupation either does not really exist or is viable only if one withdraws one's "life investment" from one's work.

The final step in the analysis—the step that makes the clearest link to organizational implications—evolved from the finding that not all occupants in a given career type react in the modal way. One asked oneself the question, therefore, of what one could learn from the "deviants"—from those whose orientation was not typical of their career type. The details of the analysis (which appear in chapter 5) will not be repeated here, but a few conclusions are worth highlighting.

In the scientific-professional occupations the congruence of organizational rewards, orientation, and job satisfaction is generally high. Most of these people are technically oriented, and this orientation was found to be rewarded by the employing organization and satisfactory to the employee. In contrast, the engineering-based occupations show some lack of congruence, and such divergences tell us something about the strains in these occupations. In high-level engineering-based occupations the technically oriented alumni are most satisfied with their careers, despite the fact that an orientation toward people is more consistent with the requirements of these jobs and is the modal orientation. In the lower-level engineering-based occupations job and career satisfaction are generally low, but the few people who do show higher satisfaction are not the ones one might expect. In those jobs it is the technically oriented people with below-average relative incomes who are most likely to be very satisfied. Jobs at this level are evidently not meeting the needs of the ablest engineering-educated people whose orientations continue to be primarily technical.

Implications for the Organization

In considering the implications of the findings, we concentrate less on the alumni who are consultants, independent professionals, or professors, and emphasize the larger group who are employed in staff and management positions and who are subject to the managerial policies of their employing organizations. This section suggests ways in which organizations could think more creatively about the various kinds of people carrying on careers within them, with the goal of providing the possibility of career satisfaction for all and thus increasing the effectiveness of the technical, professional, and managerial work force.

The starting point is the finding that people have different orientations toward their work at mid-life, even those whose educations were similar. At mid-career some alumni are oriented toward the technical aspects of work, some more toward the human aspects, and some are oriented away from work altogether and are more concerned with community, family, and other noncareer kinds of activities. Organizations must be aware of these differences, but must not equate different orientations with differences in overall ability and effectiveness. The data on relative income (which show that employees in each of these orientations have incomes above the means of their occupational/age groups) support the contention that effective performance is possible for each orientation. In the management of technical personnel, therefore, employees' abilities and the basic orientations of their lives must both be taken into account. Each combination of ability and orientation can contribute to an organization in different ways, and each requires different organizational responses. Managerial strategies must be based on an understanding of the psychological imperatives that shape career patterns as well as on an accurate measure of an employee's skills.

The key to this way of looking at the issues is the matrix that results from combining each of the three orientations discussed above with two levels of effectiveness and potential. It is presented in chart 7.1, which indicates some of the organizational roles particularly suitable for each combination.[1] It must be stressed that the six "cells" in chart 7.1 represent six quite different kinds of psychological contracts between employers and their employees, and that all of them represent legitimate and necessary roles in an effectively functioning organization. But the organization and the individual must recognize the reality on which the psychological contract is based in each case, and must generate incentives and rewards congruent with this reality.

Cell T+ This cell represents the technically oriented alumni who are viewed as high performers by their employing organizations. This group would include techni-

Chart 7.1 Midcareer Organizational Roles

Orientation at Mid-Career	Organizational Evaluation of Effectiveness High (+)	Low ("ordinary") (−)
Technical (T)	**T+** Independent contributor Policy specialist "Idea innovator" "Internal entrepreneur"	**T−** Technical support Expert on "formatted" tasks "Master"
Human (H)	**H+** Top management Sponsor Development as policy "Successful" manager	**H−** Mentor Individual development functions "Coach" "Effective" manager
Non-Work (NW)	**NW+** Specialist Internal consultant "Variance sensor" "Scanner"	**NW−** Routine tasks Reduced time commitment

cal project managers, independent contributors of various kinds, and others whose technical orientation inclines them to retain a technical core to their careers. In it one would find the "idea innovators" and "internal entrepreneurs" identified by Thompson and Dalton (1976). It is this group for which the "dual ladder" was originally invented, though the application of the concept has usually been limited to a very small group of research-and-development specialists or wrongly applied to certain managers who are no longer seen as making valuable contributions.

A critical problem for the management of employees in this cell is finding a way for them to participate in company decisionmaking, particularly on technical matters which concern them most. The data from this study have indicated that a number of people in this cell move into management because they see this as the only way to contribute to the policy decisions that affect them. Such a move is probably not an optimal solution in the long run. However, in some cases temporary assignments of these people to H+ positions might be of use. Such a strategy would assure the high-potential technically-oriented employees a central role in the organization, and would also periodically test their orientation and help them and the organization see whether it continues to be technical. It is a valuable strategy also because it provides organizational flexibility to set up temporary project-management structures, which are effective for the solution of certain complex technical problems or the implementation of major technological innovations.

In general, if these people are to be provided with viable lifelong careers, and if their technical orientation remains stable, organizations will have to develop reward systems that are congruent with this orientation. One might imagine more emphasis on recognition for technical accomplishments, more sharing of patent rights, more encouragement of attendance at professional meetings, and more use of sabbaticals and company-supported educational efforts—rewards that are specifically meaningful to technically oriented people. The point is made by a 1959 mechanical engineering graduate:

Industry in general tries to squash out every instance of creativity and ingenuity it can, due to its fear of change. A perfect example is the present patent assignment policy of most companies. Changes should be encouraged allowing the engineer to enjoy some of the fruits of his own labors.

A particularly dramatic (and, one hopes, rare) example of mismanagement in this cell is reported by a 40-year-old chemical engineer:

I have had to write papers and sections of books which appeared under the authorship of my supervisor three levels up, on matters he can hardly understand, much less contribute to except by proof reading for grammatical errors . . . The four key people whose work he became a world-recognized success by are disposed of as follows:

(1) Dead, heart attack, age 53, Ph.D. Chemical Engineering
(2) Dead, heart attack, age 42, M.S. Chemistry
(3) Dismissed from his job, age 49, Ph.D. Chemistry
(4) Mental breakdown, 2 months in psychiatric hospital, age 36, Ph.D. Chemical Engineering, currently seeking other employment.

Cell T− This cell consists of those technically oriented alumni who are perceived as average in abilities or potential by their organizations. It is a group that probably includes the steady, solid contributors to the organization, the ones who do the bulk of the day-to-day technical work. Here one finds the midlife technologists who are still technically challenged by their work, even though they are probably not able to function well either as managers or as high-level technical contributors. How to keep such individuals from losing the work motivation they have poses a real challenge to managers. The obvious response to this challenge, which lies in organizing work for the specific purpose of keeping it technically challenging and involving, runs against some accepted organizational procedures. Because many of the people in this cell may be technically less up to date than recent graduates from technical institutions, and because younger employees are less expensive, organizations tend to be biased toward assigning the new and interesting projects to the recent graduate rather than to the more experienced but less formally up-to-date employee. Though such a strategy may make short-run financial sense, it may create serious long-run problems in the management of the technical work force.

First, it is likely to reduce the work orientation of mid-career employees and effectively move them into cell NW− One way to counteract this might be to pair employees in this cell with those in cell T+ as a way of integrating technical skills with policy issues in attacking problems. Cell T− people at mid-life would probably play this role better than their young counterparts, who are still too involved in learning how the organization works and figuring out how to get into the top ranks themselves. Second, evidence from studies of newly hired technical employees shows that their academic knowledge is not sufficient for effective performance in a business organization (Jacobson 1977). They must learn to integrate this knowledge with the needs and circumstances of the particular organization in which they are employed. The experienced T− employee could provide help here—an apprentice-master pairing of young and "old" in this cell might well serve the needs of both. Finally, obsolescence—particularly that which deals with knowledge in a technical specialty (called "areal" obsolescence by Ferdinand 1966)—may be alleviated by proper concern for work assignments. There is increasing evidence that midcareer technologists bring themselves up to date when given a challenging assignment that requires new knowledge or skills (Dalton and Thompson 1971).

It seems, therefore, that by ignoring the needs and experience of midcareer employees in this cell, organizations are underutilizing them and may be creating a group of disaffected employees where none need exist.

Cell H+ This cell represents alumni with a human orientation who are considered particularly effective by the organization, and should include most of an organization's general managers as well as those who play key roles as project leaders, sponsors, and task-force chairmen and those who serve in other organizational roles that require both interest and competence in managing people. In this group one is likely to find the future top executives of a company.

The incentives of high income and promotion typically available in organizations seem to fit well with the career needs of H+ employees. In fact, human resource planning, reward systems, and performance appraisals are typically designed by higher-level people with this group of employees in mind. There is a danger, therefore, that all middle employees in an organization will be perceived in terms of the requirements of the H+ cell. Those who fit will be judged successful by the organization; those who do not fit will be perceived as lacking in ambition or as too rigid in orientation. Senior managers often tend to consider non-H+ employees unworthy of much attention. The argument here, in contrast, is that employees in all of the cells contribute to organizational success and must be provided with meaningful organizational incentives and rewards.

Cell H− In this cell are employees with a human orientation who are not seen as having the potential to rise in the organization. These include employees who never were very technically oriented as well as those who have become more oriented toward people in mid-career but are assumed not to possess the specific talents necessary to progress into top managerial jobs. These are the people most likely to be described as "plateaued" or "dead wood"—victims, perhaps, of the Peter Principle. They are perceived as not contributing significantly either to technical or to managerial tasks. Assignments that draw on their technical background and experience but also involve working with people should be explored for this group. There is some evidence, for instance, that technical careers are enhanced by mentors (Thompson and Dalton 1976) and that being a mentor is a satisfying midlife stance (Levinson et al. 1978). Such a relationship requires from the mentor much greater personal involvement than is usually the case when a top executive sponsors the person seen as most likely to succeed to that top job. Midlife H− employees seem to be uniquely qualified to play these roles, particularly for those employees—the majority in a company—who are not moving to the top. H− employees could be the "effective" if not the "successful" managers (Graves 1978), and thus could be important in the development of organizations. Further, as Rhoades et al. (1978) note, the innovation process in organizations involves a multiplicity of technical and interpersonal roles, and H− people might be particularly suited to the interpersonal function ("coaching") of project management.

Cell NW+ This cell, which represents the organization's major lost resource, consists of high-potential people who have lost their work orientation or who never had very high work motivation in the first place. Within this group are both human and technically oriented individuals who are capable of major managerial or technical contributions but, for one reason or another, are more involved with activities outside the organization, such as family, hobbies, or community. Almost by definition, these are the employees least likely to follow the "rules of the game." The rewards and incentives associated with traditional organizational career paths are not likely to be effective here, since these tend to be rigid and unresponsive to "deviant" needs. Thus, employing organizations that do not want to lose the contribution of these capable people must be willing to be flexible on work demands and to negotiate special roles for them. One person of this type might want to have a consulting arrangement with the company, working intensively at certain times on certain problems and then withdrawing for a while. For another, a temporary assignment to an H+ job might be possible, if paired with a sabbatical. Bennis (1976) suggests roles necessary in modern organizations (the "variance sensor" and the "scanner") that might well be filled by NW+ people, since they are likely to have the necessary distance from the organization to couple an outsider's perspective with an insider's knowledge.

In general, the optimal utilization of NW+ employees acknowledges their priorities and thus requires innovative arrangements between the organization and the employee based on individual negotiation. It would seem wise to give thought to these people, who are likely to become more numerous as the younger generations get to the middle years.

Cell NW– Here are found average employees not oriented to their work. Sometimes such a non-work orientation develops when employees are forced into organizational roles incongruent with their interests and capacities (Gordon 1977); at other times it evolves as a response to unfulfilled hopes of promotion. There are times, however, when such an orientation is not a reaction to adverse work experiences, but represents basic values and commitments—a situation that may be becoming more prevalent.

How an organization can best respond to the problems of its NW– employees will depend partly on the reasons behind their non-work orientations. It will also depend on the organization's rate of growth, the environment in which it is operating, and its overall policy toward its employees. A rapidly expanding company in a period of general economic growth may be able to dismiss all its NW– employees. In contrast, a more paternalistic organization or a regulated industry in a region with high unemployment may find it more expedient and less costly in the long run to develop these employees and make the effort to find roles for them that fit their motivations and capacities.

The typical managerial response to this group is that they must be "remotivated,"—an application of the H+ managerial stereotype that low work involvement is bad. But low involvement does not automatically mean that work is poor. Indeed, many tasks in organizations are pretty routine and might be best handled by less involved employees. Also, this group might respond well to opportunities for part-time work, job sharing, or various other arrangements for time off (see Cohen and Gadon 1978).

Managers must recognize that efforts to "motivate" NW− employees may come to naught, but that high levels of work involvement are not necessary for all members of an organization and may be dysfunctional for some.

Finally, if the other types of employees are handled properly, the NW− group is likely to be small. As already indicated, it is often organizational policies rather than "bad" employees that augment the size of this group. Organizations may respond to work-oriented (whether technically or human-centered) employees of average capacity in such a way that their work orientation disappears, or they may not provide NW+ employees with the conditions under which their potential can be expressed and thus effectively push them into cell NW−. In other words, by not looking at the varying orientations of mid-career people, and by not dealing as intelligently with the many employees of ordinary effectiveness as with their most able personnel, organizations unnecessarily increase the size of this group.

Concluding Note

The findings of this study show that despite common educational backgrounds, despite similar early-career interests and experiences, and even despite consensus on job values, people at mid-career have a variety of orientations to their work. Living with technology is not easy, and the technically trained find many ways of responding to the dilemmas posed. Indeed, some shift the whole focus of their lives to areas other than work.

The successful pursuit and management of technical careers requires an appreciation of the whole range of possible orientations. It requires, also, an accurate assessment of performance, an assessment on which the employee and the employing organization must concur. Most important, the variations in orientation and in ability necessitate flexible organizational policies and multiple organizational roles. Only if these exist can organizations make the most of the contributions of their technical personnel and create an environment in which technology-based careers present long-lasting satisfactions to those who pursue them.

When one adds to this conclusion the implications of the findings on the women graduates, one begins to get a sense of emerging issues in the management of human resources. Those data support the possibility of satisfactory and effective careers without full involvement. They also indicate the dangers of assuming that there is only one way successfully to traverse the early years of a career. In particular, they imply that technical careers can survive slow starts, late entrances, and midcourse changes in direction, and that over a lifetime the individual may even benefit from such unorthodoxies.

In the near future, organizations will be confronted increasingly with employees seeking a variety of career paths which move across and down as well as up and which cross organizational boundaries with ease. Such demands will require innovative and flexible responses. One of the great challenges that lie ahead for management is the acknowledgment and creative use of this pluralism.

Notes to Chapter 7

1. In the section the discussion draws on general knowledge of organizations as well as on the results of this research. These organizational implications are informed by the data but are not proved by them. It is to future research based on actual changes in organizational procedures that one must look for full validation of these conclusions.

Appendixes

A Supplementary Tables

All tables in the appendixes except table B.1 refer to men only; table B.1 refers to men and women.

Table A.1 Occupational Roles by Year of Graduation

	1951		1955		1959	
	Number	Percentage	Number	Percentage	Number	Percentage
Entrepreneur	40	7%	27	7%	15	3%
General manager	26	5%	15	4%	9	2%
Functional manager	75	14%	50	13%	32	7%
Science manager	11	2%	5	1%	4	1%
Engineering manager	99	19%	70	19%	63	14%
Staff scientist	20	4%	14	4%	26	6%
Staff engineer	114	21%	69	19%	123	28%
Business staff	29	5%	20	5%	21	5%
Science professor	13	2%	17	5%	39	9%
Engineering professor	16	3%	16	4%	30	7%
Consultant	21	4%	17	5%	22	5%
Total	464	87%[a]	320	86%	384	86%[a]
Total no. of respondents	534	100%	371	100%	446	100%

a. Percentages do not always add up because of rounding errors.

Table A.2 Some Background and Current Characteristics of Respondents in Eleven Occupational Roles

	Entr. (N=82)[a]	Gen. Mgr. (N=50)	Funct. Mgr. (N=157)	Sci. Mgr. (N=20)
	Background Characteristics			
Father's Occupational Status				
Professional				
top	6%	8%	6%	0
lower	15%	20%	15%	35%
Business				
major	26%	41%	21%	10%
small	28%	16%	22%	20%
Lower white collar	14%	8%	24%	25%
Blue collar	11%	6%	12%	10%
	Current Employment Characteristics			
Employment Sector				
Private	100%	100%	92%	65%
Nonprofit	0	0	4%	10%
Government	0	0	4%	25%
Income (1970 Dollars)				
≤ $15,000	5%	0	4%	10%
$15,001–$20,000	8%	4%	28%	15%
$20,001–$30,000	36%	26%	48%	55%
> $30,000	51%	69%	19%	20%
	Current Family Characteristics			
Family Status				
Single	4%	0	7%	6%
Married, no children	4%	2%	1%	11%
Married, preschool children	32%	41%	37%	33%
Married, school-age children	60%	56%	54%	50%
Wife's Professional Status				
Nonprofessional	77%	88%	79%	79%
Semiprofessional	10%	8%	16%	16%
Full professional	13%	4%	5%	5%
Wife's Work Status				
Not working	73%	86%	77%	74%
Working part time	19%	8%	21%	16%
Working full time	8%	6%	2%	10%

a. N values are reduced by the number of people about whom information on any category is not available.

Eng. Mgr. (N=232)	St. Sci. (N=60)	St. Eng. (N=306)	Bus. St. (N=70)	Sci. Prof. (N=69)	Eng. Prof. (N=62)	Cons. (N=60)
7%	14%	4%	8%	18%	7%	7%
17%	32%	20%	16%	15%	26%	15%
14%	17%	13%	22%	13%	13%	18%
26%	15%	19%	25%	19%	23%	23%
16%	8%	22%	18%	13%	16%	23%
20%	14%	21%	10%	21%	15%	13%
81%	48%	84%	79%	0	0	100%
10%	30%	9%	7%	100%	100%	0
10%	22%	8%	14%	0	0	0
4%	18%	20%	16%	42%	19%	9%
27%	46%	52%	56%	33%	29%	18%
58%	36%	27%	25%	24%	43%	38%
12%	0	1%	3%	0	9%	34%
6%	7%	11%	2%	11%	14%	10%
5%	11%	7%	7%	18%	10%	9%
41%	41%	40%	46%	46%	39%	38%
48%	41%	42%	46%	26%	37%	43%
79%	61%	78%	75%	54%	66%	74%
15%	26%	16%	18%	31%	17%	18%
6%	13%	6%	7%	15%	17%	7%
74%	67%	74%	76%	49%	63%	79%
20%	18%	18%	15%	26%	20%	13%
6%	15%	8%	9%	25%	17%	8%

Table A.3 Percentage of "Successful" Respondents Rating Other[a] Job Characteristics "Very Important"

	Pattern E: Engineering-Based			
	Total 314[b]	Entr. (N=42)	Gen. Mgr. (N=39)	Funct. Mgr. (N=53)
Recognition	32%	23%	24%	38%
Sufficient family time	18%	15%	14%	15%
Location	21%	18%	27%	17%
Friendly department	15%	13%	8%	11%
Reasonable work load	18%	18%	18%	17%
Contribution to society	16%	22%	27%	9%
Highly regarded organization	18%	18%	22%	17%
Work with people	16%	15%	40%	15%
Job with prestige	18%	31%	19%	21%
Efficient department	13%	18%	14%	14%
Job security	7%	0	3%	11%
Social relevance	5%	2%	3%	4%
Fringe benefits	6%	5%	8%	8%
Physical working conditions	4%	0	10%	8%

a. Compare table 3.3.

b. N values represent the number of people in each category who consider themselves "successful"; they are reduced where necessary by the number not answering a given item.

				Pattern SP: Scientific-Professional				
Eng. Mgr. (N=75)	Cons. (N=34)	Bus. St. (N=20)	St. Eng. (N=51)	Total 71	Sci. Mgr. (N=12)	Eng. Prof. (N=25)	Sci. Prof. (N=17)	St. Sci. (N=17)
30%	30%	40%	41%	34%	36%	48%	12%	35%
15%	21%	10%	29%	13%	17%	22%	12%	0
17%	15%	20%	31%	16%	0	29%	12%	12%
17%	21%	15%	20%	12%	8%	22%	12%	0
17%	21%	5%	22%	16%	17%	14%	25%	12%
12%	21%	15%	14%	22%	0	44%	6%	24%
17%	18%	15%	16%	17%	17%	26%	12%	12%
9%	18%	20%	8%	6%	17%	9%	6%	6%
15%	9%	20%	16%	17%	17%	17%	18%	18%
12%	15%	15%	10%	9%	17%	4%	12%	6%
7%	3%	0	18%	16%	0	18%	29%	12%
4%	15%	5%	6%	9%	8%	22%	0	0
7%	3%	5%	6%	6%	17%	9%	12%	0
3%	6%	0	2%	3%	33%	4%	6%	0

B Population and Sample by Year of Graduation

Though there are some discernible age trends in the sample (Bailyn and Schein 1972), the differences among the three classes are relatively small. Throughout most of this book, therefore, the classes are combined. Data given separately for each class include the distribution of undergraduate majors in this appendix and the occupational distribution of the male alumni in table A.1..

Table B.1 shows the distribution of undergraduate majors in the sample and the population of each class. The population figures given in the last line of the table represent the number of questionnaires actually sent out. This number is somewhat less than the total number of graduates in each year because some alumni's addresses were not available at MIT. In general, there was a 7%–8% loss of this kind for each class.

Table B.1 shows, in the first two columns for each class, the number and percentage in each undergraduate field who responded to the questionnaire. A comparison of this information with the percentages in the third column (representing the percentage of the population as given by the number of questionnaires sent out) indicates that, at least as far as undergraduate major is concerned, there is no bias in the sample.

It is evident also that the distribution of majors in the class of 1955 is almost identical to that in the class of 1951. But the class of 1959 shows a trend away from engineering and towards science. This trend became much more marked in the years that followed. Only 43% of the graduates from the class of 1977 received degrees in engineering, while 39% were in science. But even though these changes are anticipated in the youngest class studied, the differences are still so small that it seems entirely justified to treat the three classes as a unit.

Table B.1 Representation of Undergraduate Majors in Population and Sample, by Class

Undergraduate Major	Class of 1951 Sample No.	%	Pop. %
School of Engineering			
Civil Eng.	37	7%	6%
Mechanical Eng.	87 (1)[a]	16%	15%
Metallurgy	27	5%	4%
Electrical Eng.	73	14%	16%
General Eng.	10 (1)	2%	4%
Chemical Eng.	64 (1)	12%	12%
Marine Eng.	10	2%	2%
Aero. and Astro.	32	6%	5%
Construction Eng.	19	4%	4%
Total	359 (3)	67%[c]	68%
School of Science			
Chemistry	14	3%	3%
Biology	5	1%	1%
Physics	30	6%	5%
Geology	13	2%	2%
Mathematics	5	1%	1%
Meteorology	6	1%	1%
Nutrition	3 (1)	1%	1%
Total	76 (1)	14%[c]	14%
School of Management	68	13%	12%
School of Architecture and Planning	23 (1)	4%	4%
School of Humanities and Social Science			
Economics	13	2%	2%
Humanities	0[d]	0	[n=0]
Total	13	2%	2%
Total No.	539 (5)		883 (6)

a. Numbers in parentheses indicate number of women.
b. Discontinued after class of 1957.
c. Percentages do not always sum exactly because of rounding errors.
d. This major was only approved by the faculty in January 1955, and only available to the class of 1958 and succeeding classes.

Class of 1955 Sample			Class of 1959 Sample		
No.	%	Pop. %	No.	%	Pop. %
20	5%	7%	23	5%	5%
54	14%	13%	68	15%	14%
16	4%	4%	29	6%	5%
71 (2)	19%	22%	86 (1)	19%	21%
14	4%	4%	1	0.2%	[n=2]
44 (1)	11%	10%	34	8%	9%
5	1%	2%	5	1%	1%
27	7%	6%	23	5%	5%
6	2%	1%	0[b]	0	[n=0]
257 (3)	68%[c]	68%[c]	269 (1)	60%[c]	60%
8	2%	2%	13 (1)	3%	3%
3	1%	1%	11	2%	2%
24 (2)	6%	7%	41	9%	11%
6	2%	2%	7	2%	1%
7	2%	2%	19 (1)	4%	4%
0	0	[n=1]	0	0	[n=1]
4	1%	1%	3	1%	1%
52 (2)	14%	14%[c]	94 (2)	21%	22%
49	13%	13%	56	12%	11%
19 (1)	5%	5%	10	2%	3%
0	0	[n=2]	12	3%	2%
0[d]	0	[n=0]	9 (1)	2%	2%
0	0	[n=2]	21 (1)	5%	4%
377 (6)		614 (10)	450 (4)		726 (6)

C Occupational Success

Measures of career success may be based on objective or subjective criteria. Objective measures—based on such indicators as income, rank, and output—have certain advantages of rigor, but they also produce some problems. Though their objectivity may make them preferable when studying differences in success within a homogeneous professional group, occupational differences in the distribution of such measures make them potentially meaningless if a sample spans many different occupations. In contrast, subjective measures—based on some form of self-rating—tend to overcome such occupational differences, since presumably each person is rating himself relative to his own occupational group. This method also has problems. The match between objective and subjective measures will be greater in convergent fields, where criteria of excellence are much more specified, than in divergent ones. Further, in any given field, biases are introduced by the dependence of self-ratings on a person's degree of self-awareness and the strength of his needs and defenses. On the other hand, subjective measures may have a further, more subtle advantage in allowing each person to use his own particular criteria of success. Some of the respondents mentioned that they did not consider success in their work relevant to success in life. It seems likely that they were interpreting success in terms of objective criteria (income, possessions, etc.). Subjective ratings allow these people to include other, more personally relevant aspects of their work, such as the amount of satisfaction they have from it and the level of ability at which they are working.

It is for this reason that the analysis used a primarily subjective measure of success, centered on item 19 in the questionnaire: "At this point in your professional life, how successful do you think you are in your work?" A difficulty concerns the effect of career stage on answers to this question. Since there are three somewhat different age groups in the sample, and since career lines differ in different occupations, such a measure is probably not as good as one that also takes into account a person's estimate of his success at the height of his career, whenever that may be. We therefore included one further question

Table C.1 Occupational Success as Measured by Combination of Responses to the Two Questions of Item 19

	Success at Height of Career					
	Low	Medium	High	Total No.	% Expecting Same Success	% Expecting Improvement
Success Now						
Low (1–3)	120	253	74	447	27%	73% (17% to High)
Medium (4)	9	407	243	659	62%	37%
High (5)	2	21	185	208	89%	—

in item 19: "How successful do you think you will be at the height of your career?"

Occupational success, then, was measured by a combination of these two questions, as presented in table C.1. That table shows that essentially all of those rating themselves as very successful now also do so for the height of their career (185 out of 208); about one-third of the current "mediums" feel they will move up (243 out of 659); and almost three-fourths of those currently rating themselves as relatively unsuccessful expect to move up (327 out of 447), though most only to the medium position.

We consider "successful" those respondents who fall below the dashed line in table C.1 (those who rate themselves as very successful now and those who rate their present success as medium but expect to be very successful at the height of their career). About one-third (N=451, 34%) of the male sample who answered these questions are "successful" by this measure.

Table C.2, which presents perceived success in relation to career patterns and occupational roles, shows that there is no difference in perceived success between the career patterns. There is considerable variation, however, among the occupations within these patterns. Over three-fourths of the general managers perceive themselves as successful—the highest "success" of all the groups. They are followed by three occupational roles in which more than half perceive themselves as successful: the science managers, the consultants, and the entrepreneurs. Next come a group of occupations in which about one-third perceive themselves as successful. These are, in descending order, the engineering professors, the functional managers, the engineering managers, business staff, the staff scientists, and the science professors. (The lower felt suc-

Table C.2 Perceived Success in Career Patterns and Occupational Roles

	% Who Perceive Themselves Successful[a]	Rank[b]	Correlation[c]
Pattern E	**33% (N=947)**		**0.19**
Entrepreneurs	52% (N=80)	4 (2)	0.40
General managers	78% (N=50)	1 (1)	0.28
Functional managers	34% (N=155)	6 (4)	0.19
Business staff	29% (N=70)	8 (8)	0.21
Consultants	58% (N=59)	3 (3)	0.25
Engineering managers	32% (N=231)	7 (7)	0.16
Staff engineers	17% (N=302)	11 (10)	0.22
Pattern SP	**33% (N=211)**		**0.17**
Engineering professors	40% (N=62)	5 (6)	0.18
Science professors	25% (N=69)	10 (11)	0.13
Science managers	60% (N=20)	2 (5)	0.48
Staff scientists	28% (N=60)	9 (9)	0.18

a. Includes only respondents who answered both success questions.
b. Ranks in parentheses are based on the percentage in each group whose 1970 incomes were over $30,000, with ties allocated according to the percentage in the $20–$30,000 range. In order to control for career stage, these rankings were done separately for each class and the final ranks given in this column are based on the average ranking of the 3 classes for each occupational group.
c. These are nonparametric correlations between perceived success and income normalized within each occupational category and graduating class.

cess of the science professors in comparison with the engineering professors may well result from the tendency among scientists to measure their efforts against those of the giants in their fields [Kubie 1953, 1954].) Finally, very much at the bottom are those respondents in staff engineering positions, of whom only 17% perceive themselves as successful.

Table C.2 also ranks the occupational groups according to the percentage reporting very high incomes. A comparison of this ranking with the subjective ranking shows a strong but not perfect correlation (rank order correlation = 0.91). For example, the entrepreneurs and the functional managers (in the E pattern) perceive themselves as less successful than their incomes indicate, whereas alumni in the SP pattern, particularly the science managers, tend to see themselves as more successful than their incomes show. This overall relation, however, applies to groups and does not say anything about the correlation between perceived success and income for individuals within the occupational categories. The last column of table C.2 provides this information. Though all but one of these correlations (the one for science faculty) is statistically significantly different from 0 ($P<0.05$), only for the entrepreneurs and the science managers are the correlations high enough to warrant the conclusion that perceived success depends to any extent on relative income.

D Work Involvement

The index of work involvement emerged from a factor analysis of 24 variables, some of which were based on single items in the questionnaire and some of which were based on scales or item combinations. The entire array was factor-analyzed by means of a principal-factor solution with iteration (Nie et al. 1975). Six factors with eigenvalues >1.0 which met the scree test (Gorsuch 1974) were rotated by both an orthogonal (varimax) and an oblique (direct oblimin with Kaiser normalization) rotation. In each case there emerged a factor (the first one after the varimax rotation) on which items previously identified as a work-commitment scale loaded simply and highly. This factor served as the basis for the index of work involvement. Since the oblique rotation presented a clearer picture of all six factors, the following results are based on it.

Six of the original 24 variables had loadings greater than 0.3 on the "work involvement" factor, and five of these loaded higher on this factor than on any other. Of these five, one had a considerably lower loading than the others (0.32); it also had an almost equal loading (0.28) on another factor. For this reason it was eliminated from the definition of work involvement, leaving the four variables shown in table D.1 to be used in the final definition.

The first variable, career satisfaction, is based on the ranking given to "career or occupation" in the response to item 31, which asked each respondent to select and rank which three of the following aspects of life give the most satisfaction: career or occupation, creative or other activities not related to career or occupation, leisure-time recreational activities, family relationships, and activities directed at community, national, or international betterment.[1] This question was designed to compare career

Table D.1 Final Variables Used in Index of Work Involvement

	Loading
Career satisfaction	0.71
Work satisfaction	0.64
Importance of time for family and personal life	0.61
Work orientation	0.56

satisfaction with satisfaction in these other areas. Only family ranks anywhere near career as providing the most satisfaction in life for this sample: 45% ranked career first; 40% ranked family first. And 64% mentioned family and career as the two most important sources of satisfaction (of these, 51% put family first and 49% put career first).

The second variable listed in table D.1 also deals with satisfaction. It represents the extent of agreement with the last statement in item 26: "My main satisfactions in life come from the work I do." The mean ranking of the 5-point response scale—from 1 ("strongly disagree") to 5 ("strongly agree")—was 3.0: 7% agreed strongly and 26% responded "4." Thus, about one-third of the respondents agreed with this statement. A somewhat smaller number disagreed: 8% responded "1" and 19% answered with a "2." (Neither of these first two variables had a loading even as high as 0.1 on any other factor.)

This second variable is one of five parts of a question taken from a "commitment-alienation" scale used by Rapoport and Lohman in their study of physical scientists and engineers from three different technological institutions 10 years after graduation. It was not combined with the other parts of that scale because we considered it possible that this question along with the question on career satisfaction and one asking "How satisfied are you with your present job?" might constitute a "satisfaction" factor. But no such cluster emerged.[2] The first item in this scale, the extent to which a person would feel "very frustrated and unfulfilled" if he "had to change the kind of work I do," was also kept separate, because it gave consistent results for some occupations but not for others. The crucial point has to do with an ambiguity in the concept of "work." Scientists, for example, seem to see work and career as synonymous, and hence the scale produced consistent results for them. In contrast, a number of entrepreneurs responded in the direction of involvement on all the other parts but said on this item that they would not mind changing the kind of work they did (implying, perhaps, that their concept of career is not dependent on the exact nature of their daily tasks).

The third variable in the "work involvement" factor stems from the question on job values (item 17). It asks for the importance assigned to the fact that a job "leaves sufficient time for family and personal life." On a 5-point scale from 1 ("not at all important") to 5 ("very important"), the mean response was 3.5; 21% responded "5" and 29% gave a rating of "4."

The last variable included in the definition of work involvement is that of work orientation. It represents an average response to the three middle parts of item 26 on the questionnaire: "I like to think about my work, even when off the job," "My only interest in my job is to get enough money to do the other things that I want to do" (reverse scoring), and "I wish that I were in a completely different occupation" (reverse scoring). The mean score on this variable is 4.1 (on a range from 1.0 to 5.0), which indicates once again how little the work motivation of the respondents centers solely on instrumental concerns. In fact, over 50% of the group disagreed strongly (response 1) with statements 2 and 3. Somewhat fewer (36%) were in strong agreement with the first statement, but here too around three-fourths of the respondents came out on the "committed" end (responses 4 and 5 to the first statement, 1 and 2 to the second and third).

These four variables were converted into three equally weighted elements that constitute the index of work involvement. The first element is measured by combining the two variables that stem from item 26, work satisfaction and work orientation, into one scale ranging from 1 to 5. In the previous study in which this question was used, all five parts belonged to one scale. They were entered as three separate variables into the original array only because there was some doubt whether they would scale in the present sample. This doubt was confirmed for the statement concerning frustration at a hypothetical change in one's work. Since, however, the factor analysis showed that the last item (work satisfaction) loaded on the same factor as the middle three (work orientation), it seemed most reasonable (indeed methodologically advisable, since the two variables must share a fair amount of "method" variance) to recombine them. The second element is measured by the question on career satisfaction already described. However, instead of taking the ranking of "career and occupation" as the indicator, the implicit comparison of career and family was made explicit by adjusting the rank given to career according to that attrib-

Table D.2 Distribution on Final Measure of Work Involvement, Based on Average of Three 5-Point Scales

Extent of Work Involvement	Average Score	Percentage[a]	
Very high	4.0–5.0	9%	25% High
High	3.6–3.9	16%	
High medium	3.1–3.5	25%	52% Medium
Low medium	2.5–3.0	27%	
Low	2.0–2.4	14%	23% Low
Very low	1.0–1.9	9%	

a. Percentage of the 1,270 respondents who answered all component questions.

uted to "family relationships."[3] The last element is measured by reversing the scoring on the degree of importance (on a 5-point scale) given to time for family and personal life.

The final measure, then, consists of the average of these three 5-point scales,[4] and results in the distribution shown in table D.2 (with a mean score of 3.02).

Table D.3 shows the degree of correlation between the total index and its various components. Though the total index correlates more with the family-related items than with the work-commitment items, the fact that it has a higher correlation with the career-satisfaction part of the first component than with the family-satisfaction part seems to indicate that career and family play fairly equal roles in the index. The separate parts of the work-commitment component yield some interesting information. One may think of parts (a) and (b) as two sides of the intrinsic-instrumental coin. Agreement with part (a) indicates an intrinsic orientation, agreement with part (b) an instrumental one. In a similar vein, parts (c) and (d) may be viewed, respectively, as indicators of dissatisfaction and satisfaction with work. Looked at in this way, the correlations indicate that the index is more closely related to the intrinsic, satisfaction components than to the instrumental, dissatisfaction ones.[5]

This index, therefore, reflects a combination of feelings and attitudes that hang together in the respondents' views of their careers and lives—a syndrome of reactions to work and to the role that work and career play in one's total life.

Table D.3 Correlations of Work-Involvement Index with Components

	Correlation	
Major life satisfaction		0.81
(a) Rank for career	0.74	
(b) Rank for family	−0.62	
Importance of time for family as job characteristic		−0.75
Commitment to work		0.68
(a) Think about work when off job	0.51	
(b) Only interest in job is money	−0.37	
(c) Wish were in different occupation	−0.39	
(d) Satisfaction from work	0.60	

Explication: Extreme Groups

Some of the impressionistic distinctions described in chapter 4 between the 25 most highly work-involved respondents and the 23 with the lowest scores on this index are caught in the numerical differences shown in table D.4. This table shows that some of the respondents with extremely low scores on work involvement give spontaneous indication of uncertainty about their professional identity. Yet none of those with extremely high scores does so (lines 1–3). This difference is further highlighted by the entirely different distribution of responses to two other items from the questionnaire. The "extremely low" group is biased in their distribution of responses to the question of whether they have chosen the right career toward the "no" end, while the "extremely high" group is biased toward the "yes" end (line 4). Further, a high proportion of the "extremely low" group would not be at all frustrated if they had to change their work (line 5)—indeed, they would probably welcome it. The "extremely high" group, because of the already mentioned ambiguity on this question, responds at both extremes.

The lack of precision and realism of the "extremely low" group in their relation to their work is illustrated in the second part of the table. They seem to want everything out of their work (line 6); almost half say that 9 or more of the 22 job characteristics mentioned in the job values question are "very important" to them. In contrast, almost half of the "extremely high" group give only one or two such rankings; they seem to be able to say exactly

Table D.4 Some Differentiating Characteristics Between Groups with Extremely High and Extremely Low Work-Involvement Scores

	Work Involvement Score	
	Extremely High (≥4.4) (N=25)	Extremely Low (≤1.4) (N=23)
Professional Identity		
(1) No answer to question asking "What do you consider to be your profession?" (item 22)	0	3
(2) Spontaneous mention that they wish they were in another field	0	4
(3) Spontaneous mention that professional societies are "useless," "a waste of time"	0	3
(4) Rating of certainty of having chosen right career (item 29)		
("do not possess at all") (0)[a]	0	7
("possess to great extent") (4)	6	0
(5) Reaction if had to change kind of work done (item 26)		
not at all frustrated (1)	6	16
very frustrated (5)	4	0
Precision and Realism of Relation to Work		
(6) Mean number of job values considered "very important" (5) (item 17)	3.8	8.1
less than 3 (out of 22)	12	0
more than 8	0	10
(7) Unrealistic discrepancy[b] between present success and success at the height of career (item 19)	0	5
(8) Spontaneous mention of desire some day to start own business	2	6

a. Numbers in parentheses refer to response categories involved.

b. A realistic discrepancy is defined as one that is at most one point higher, on a 5-point scale, for height of career than at present.

what they want from their work, and presumably they are getting it. Further signs of lack of realistic planning are shown in lines 7 and 8 of the table. A number of the "extremely low" group are still looking to the future—to become successful, perhaps by starting their own business—but there is no evidence that they are actively working to bring such changes about. Their hopes for the future may rest more on dreams than on realistic plans.

Validation

The two extreme groups seem to be very different in their relation to their work. But in order to get more general validating evidence for the construct of work involvement, we turn to the question on job values, since the key conceptual issue in work involvement is whether work is viewed as merely instrumental (a means of making a living) or as intrinsically satisfying and capable of fulfilling basic motives and values.

A separate factor analysis of the question on job values extracted six factors, two of which are relevant to this point. One (the first extracted) is a measure of the importance to each respondent of extrinsic job characteristics—those not derived from the work itself but from the external conditions surrounding that work. A second deals with the importance of the intrinsic satisfactions one can derive from work. The first factor would be expected to be negatively related to work involvement, the second positively.

Combining these items into two scales—one on the importance of intrinsic characteristics (three items) and the other on extrinsic ones (six items)—confirms, of course, the "intrinsic" predilection of the sample already described. The median of the intrinsic scale is 4.67 and that of the extrinsic one is 3.17 (on ranges from 1.00 to 5.00); 38% of the respondents gave a rating of 5 to all three items on the intrinsic scale, whereas only one person answered this way to the six extrinsic items (and only 23% have an average extrinsic score of 4 or more).

But these differences are not the concern here. The concern, rather, is with the direction of the relationship of each of these sets of job values with work involvement. The Pearson correlation of the index of work involvement with the importance of extrinsic job characteristics is -0.410 ($N=1{,}229$; $P<0.001$); that with the importance of intrinsic job characteristics is $+0.134$ ($N=1{,}263$;

$P<0.001$). These correlations, though not large,[6] are obviously in the expected directions. The relationships are clearer when presented as differences in categories of responses, as in table D.5. This table shows, for both intrinsic and extrinsic characteristics, the percentage of people in each work-involvement group who fall above or below the median of each scale. It also shows the individual items that constitute each scale and the percentage in each work-involvement group that considers a particular item very important. The responses to these individual items confirm again the conclusions of chapter 3 that almost everyone in the sample values the intrinsic characteristics. As a matter of fact, the extent of agreement with the importance of the intrinsic items is so great that the utility of the measure for comparisons within this sample diminishes. But the differences that do exist, though small, relate to the index in the expected direction. This validation of the index is even greater when one looks at the extrinsic scale, which is much more normally distributed on the 1.0–5.0 range. Almost three-fourths of those whose work involvement is low are above the median on these extrinsic values, compared with less than one-fourth of those with high work involvement.

Clearly, then, the evidence from the question on job values confirms the interpretation of work involvement as a measure of intrinsic work orientation, which is negatively associated with concern about extrinsic characteristics of the working situation.

Also included in table D.5 is the one part of the original "commitment-alienation" scale that did not load uniquely high on the "work involvement" factor: "If I had to change the kind of work I do, I would be very frustrated and unfulfilled." As indicated before, it was found that respondents in some occupational roles disagreed with this statement even when there were other indications of work involvement because of a different interpretation of the word "work." But even though involvement may be accompanied by disagreement with this item, one would not expect many uninvolved respondents to agree. The last line of table D.5 corroborates this expectation: only 3% of those with low work involvement would be very frustrated if they had to change the kind of work they do,[7] as opposed to 14% of those whose involvement with work is high.

Table D.5 Work Involvement in Relation to Intrinsic and Extrinsic Job Characteristics

	Work Involvement			
	High (N=323)[a]	Medium (N=653)[a]	Low (N=294)[a]	D[b]
Importance of intrinsic characteristics (median = 4.67)				
% above median	42%	40%	27%	14.5
% below median	36%	34%	50%	
Individual intrinsic characteristics (% who feel characteristic is very important)				
Challenging work	72%	69%	53%	19
Freedom to be creative and original	56%	53%	46%	10
Work giving personal sense of accomplishment	72%	71%	65%	7
Importance of extrinsic characteristics (median = 3.17)				
% above median	23%	49%	72%	−49.5
% below median	69%	41%	19%	
Individual extrinsic characteristics (% who feel characteristic is very important)				
Location	16%	17%	30%	−14
Friendly department	11%	16%	24%	−13
Reasonable workload	8%	16%	20%	−12
Job security	6%	8%	17%	−11
Good fringe benefits	2%	4%	9%	−7
Good physical working conditions	3%	4%	6%	−3
% who would be very frustrated if they had to change the kind of work they do	14%	12%	3%	11

a. These N's are reduced by the number of people not answering a given item.

b. For individual items, *D* represents the difference in the percentages of the High and the Low groups giving a certain response. Thus, $D>0$ indicates a positive relation between work involvement and the item in question, and $D<0$ indicates a negative relation. The absolute value of *D* is a measure of the degree of association involved, constrained, of course, by the total number of people giving the particular response in question. When dealing with medians, this constraint is lifted somewhat and the *D*s allow more comparison among items and scales. Since the percentage at the median of a distribution varies from group to group, the *D* in this case is the average of that for the "above median" and the "below median" categories.

Notes to Appendix D

1. This question was adapted from similar items used by Rosenberg (1957) and Fogarty, Rapoport, and Rapoport (1971).

2. The job-satisfaction item turned out to be complex and loaded on three other factors besides the "work involvement" one. This result is consistent with findings from personal interviews with some of the respondents in this study, which showed that they can distinguish between the satisfactions derived from work (the day-to-day activities of one's occupation) and from career (the long-range occupational progression that involves changes in rank and salary, responsibility, and productivity), but that questions about one's job sometimes elicit responses dealing with work conditions of the job and at other times are answered in terms of the job's place in one's career development (David Jacobson, personal communication).

3. The following scores were assigned: For career ranked 1 or 2 and family ranked 3 or lower, currently married respondents were scored 5 and currently unmarried respondents were scored 4. For career ranked 1 and family 2, all respondents were scored 3. For career ranked 2 and family 1, all were scored 2. For career ranked 3 or lower and family 1 or 2, all were scored 1.

4. The elements as finally measured correlate with each other approximately equally:

	1	2
2	0.368	—
3	0.361	0.330

5. Wilensky (1964) showed that the determinants of alienation from work are different from those of involvement with it. See also Herzberg's (1966) two-factor theory of job satisfaction.

6. They are reduced, of course, by the homogeneity in response to some of these measures, particularly to the items of the intrinsic scale.

7. Data from the small group of people (N=38) who are not involved with their work but who indicate frustration at the prospect of having to change it corroborate the multiplicity of meanings of "work" this item seemed to elicit. Fully 66% of this group were concerned about the security of their jobs, as compared with 35% of all respondents and 48% of those with low work involvement who would not be frustrated if they had to change their work. This indicates that these people probably associated "work" more with steadiness (and hence job security) than with the tasks it entails.

E Orientations to Work

Technical Orientation

Two variables from the survey were used as indicators of technical orientation. The first is the distinction between cosmopolitan and local orientations. The concept of cosmopolitanism was first used by Merton (1949) to distinguish community leaders who are cosmopolitan from those whose orientation is local. Since then, the distinction has been found useful in many other connections. In particular, many researchers have followed the lead of Gouldner (1957, 1958) and have used it to differentiate the way professionals employed in industrial organizations react to the often conflicting pulls of professional as opposed to organizational values and standards (Shepherd 1961; Glaser 1964; Pelz and Andrews 1966). It refers to a person's responsiveness to his professional colleagues or peer group, in contrast with a local orientation, which represents responsiveness to the local authority of the employing organization.

In the present study the cosmopolitan/local distinction is based on the responses to the following two statements in item 25: "I am more concerned with how my work looks to my professional colleagues than to my boss" and "If there are conflicts between professional standards and the interests of my employer I tend to resolve them in favor of my employer." (Rapoport and Lohman used these statements in their professionalism scale.) All those who answered both questions in the professional direction ("yes" to the first, "no" to the second) are considered cosmopolitan; those who answered both in terms of their organization ("no" to the first, "yes" to the second) are considered local. The relevance of this distinction to a technical orientation is based on a previous finding (Bailyn 1977b) that those with a cosmopolitan orientation in E/L careers are considerably more involved with their work and perceive themselves as more successful in it than their locally oriented colleagues.

The second variable used to indicate technical orientation is the response to the following statement from item 26 of the questionnaire, which was shown by previous analysis to differentiate between technically and managerially oriented employees: "If I had to change the kind of

Table E.1 Technical Orientation Determined from Local/Cosmopolitan Distinction and Agreement with "Change of Work" Statement

	Strength of Agreement with "Change of Work" Statement					
	1	2	3	4	5	unascertainable
Cosmopolitan	High	High	High	Very high	Very high	Very high
Mixed	Low	Low	Medium	High	Very high	Medium
Local	Low	Low	Low	Medium	Medium	Low
Unascertainable	Low	Low	Medium	Medium	Very high	?

work I do, I would be very frustrated and unfulfilled." As mentioned, this item was originally part of a scale of work commitment, but analysis showed that only technologists responded to it on that basis. In contrast, a number of highly work-oriented managers responded that they did not care if the content of their work was changed. These two items were combined as shown in table E.1 to yield a measure of technical orientation. Only 18 out of the 1,351 male respondents fall into the unascertainable category (indicated by the question mark). The rest distribute on this orientation as follows:

High technical orientation	36%	(14% very high)
Medium technical orientation	19%	
Low technical orientation	45%	

More people in the class of 1959 have a high technical orientation than in the earlier classes: 44% in the class of 1959, 34% in the class of 1955, and 30% in the class of 1951.

Human Orientation

This orientation was "measured" by the responses to the following part of item 17 in the questionnaire: "Please circle the appropriate number to show how important you feel [opportunity to work with people rather than with things] is to you with regard to your present and future jobs." Answers range from 1 ("not at all important") to 5 ("very important"). Those who answered 4 or 5 are considered high on this orientation (with 5 representing the very high position); those who answered 1 or 2 are considered low in their orientation to people. Twenty-one male respondents did not answer the item at all, and the rest distribute in the following way:

High human orientation	38%	(12% very high)
Medium human orientation	32%	
Low human orientation	30%	

There is very little difference among the classes in human orientation: 39% in the class of 1951 are high, 39% in the class of 1955, and 34% in the class of 1959.

Non-Work Orientation

This orientation was measured by the index of work involvement described in appendix D. All those with scores ≤ 2.7 on the work involvement index are considered high on non-work orientation; those with scores ≤ 2.1 are considered very high. Those with scores ≥ 3.6 are considered low on this orientation (that is, they are considered work-oriented). The cutoff points on non-work orientation were shifted to make the percentage in this category close to that in the high categories of the other two orientations.

There were 81 male respondents who did not answer one or more of the items that constitute this index. The rest distribute as follows:

High non-work orientation	34%	(13% very high)
Medium non-work orientation	41%	
Low non-work orientation	25%	

Again, and perhaps contrary to what one might expect, there is very little difference among the classes in terms of high non-work orientation: 33% in the class of 1951 are highly non-work-oriented, 37% in the class of 1955, and 31% in the class of 1959.

As mentioned in chapter 5, the final orientations were arrived at by eliminating people who were not high on any of these orientations, or who were equally high on more than one. Thus, a person who is extremely high on human orientation is classified as oriented toward people only if he is not also extremely high on either of the other two orientations. Similarly, a person who is high, though not extremely high, on the measure of technical orientation is classified as technically oriented only if he does not fall into the high category on either of the other measures.

F The Questionnaire

Massachusetts Institute of Technology

Alumni Survey

I am requesting your help in a major study of M.I.T. alumni, designed to assess the role that an M.I.T. education plays in the lives of its graduates. This study is jointly sponsored by M.I.T.'s Office of the Undergraduate Planning Professor and the Carnegie Commission on the Future of Higher Education. The information obtained from the survey will help M.I.T. reassess its goals and educational policies and will help the Carnegie Commission make recommendations pertaining to the entire system of higher education in this country.

You will note that the questionnaire is fairly detailed and covers a number of areas. This is necessary because we are trying to obtain information about the career patterns of alumni -- on how their educational experiences, their occupation, and their family life are related to each other. Also, the large numbers involved require an objective question format, even though this sometimes forces the respondent to give his answers in terms of categories that do not permit him to express the nuances of his opinion. Please feel free to elaborate your answers in the margins or in the "Comments" section.

In order for the information obtained to be valid, maximum participation in this research project is necessary. Our pretesting indicates that it takes about 30 to 45 minutes to complete the questionnaire and I very much hope that you will be willing to give that amount of time to this important project. If for any reason you cannot fill out some portions of the questionnaire, leave those blank and fill in the rest. If you cannot fill in any of it, please send the blank questionnaire back anyway. We can only check our addresses and our pattern of non-participation if we get back all questionnaires. A stamped return envelope is enclosed for your convenience. Please return the completed, partially completed, or blank questionnaire sometime within the next two weeks.

While each questionnaire has to be numbered for purposes of analysis, your answers will be kept confidential in every way. No individual questionnaire will be seen by anyone except the immediate research staff, and only group results will be reported.

I greatly appreciate your help in this effort.

Edgar H. Schein
Undergraduate Planning Professor

EHS:lsw

Your citizenship: ______________________

1. Marital status: (circle the appropriate number and give date if applicable)
   Single . . . 1   Married . . . 2 (date: __________)   Other . . . 3 (please specify: __________)

2. Your present age: ______ Number of children: ________ Ages of children: ________________

3. Parents' and wife's education: (circle the appropriate number in each column)

	Father	Mother	Wife
Eighth grade or less	1	1	1
Some high school	2	2	2
Graduated from high school	3	3	3
Some college	4	4	4
Received Bachelors degree (or currently working for it).	5	5	5
Received Masters degree (or currently working for it).	6	6	6
Received Doctorate or equivalent (or currently working for it) .	7	7	7
Not applicable			8

4. Parents' and wife's primary occupation: (circle the appropriate number in each column)

	Father	Mother	Wife
Professional (e.g., doctor, engineer, professor)	1	1	1
Proprietor or manager (enterprise of 25 or more employees)	2	2	2
Proprietor or manager (enterprise of less than 25 employees) . . .	3	3	3
Teacher	4	4	4
Military	5	5	5
Clerical or sales	6	6	6
Service Worker	7	7	7
Skilled worker	8	8	8
Semi- or unskilled worker. .	9	9	9
Housewife.	10	10	10
Other. . . . . (please specify) ____________	11	11	11
Not applicable			12

5. Is (or was) your father in the same or related area of work as you are? (circle the appropriate number)
   Same area . . . 1   Related area . . . 2   Different area . . . 3

6. Did your mother work while you were growing up? AND, is your wife working now?

	Mother	Wife
Yes, full time in her field	1	1
Yes, full time but not in her field (or had/has no special field)	2	2
Yes, part time in her field	3	3
Yes, part time but not in her field (or had/has no special field)	4	4
No, did/does not work	5	5
Not applicable.		6

7. Your undergraduate major: ____
   If you switched majors, what was your first major? ____________________

8. Your primary residence as an undergraduate: (circle one in each row)
   Dormitory . . . 1   Fraternity . . . 2   House or apartment . . . 3
   Alone . . . 1   With parents . . . 2   With friend(s) . . . 3   With wife . . .

9. What was your undergraduate grade point average (cum)? ________ Year graduated: ________

10. Did you attend graduate school? Yes . . . 1 No . . . 2
    If yes, School: ________________ Field of Study: ______________ Degree: ________ Date: _____
    School: ________________ Field of Study: ______________ Degree: ________ Date: _____

11. ASSESSMENT OF EDUCATIONAL AND JOB EXPERIENCES: Below is a list of abilities and traits that people possess to varying degrees. In a later question we will ask you to rate yourself on these traits. Here we are concerned only with the CONTRIBUTION OF YOUR EDUCATIONAL AND JOB EXPERIENCES to the development of these abilities and traits, no matter to what extent you may possess them.

For each of the items listed, please indicate in the first column, the effect of YOUR UNDERGRADUATE EDUCATION; in the second column, indicate the effect of YOUR GRADUATE OR PROFESSIONAL EDUCATION (if you did not go to graduate or professional school leave this column blank); and in the third column, the effect of YOUR POST-EDUCATION JOB EXPERIENCES.

If your educational and job experiences contributed positively to the development of these abilities and traits, circle the +2 if they contributed a great deal and the +1 if they contributed only a little. If the extent to which you possess a given factor was not dependent on or is unrelated to your educational or job experiences, circle the 0. If a given factor was decreased as a result of your educational or job experiences, circle the -1:

EFFECT OF EDUCATION OR OF JOB EXPERIENCES			
Decreased ability or trait	Did not change or not relevant	Increased ability or trait	
-1	0	+1	+2
		a little	a great deal

EFFECT OF:	undergraduate education				graduate education				post-education job experiences			
Knowledge and Abilities												
Real competence in your chosen field	-1	0	+1	+2	-1	0	+1	+2	-1	0	+1	+2
Ability to identify problems	-1	0	+1	+2	-1	0	+1	+2	-1	0	+1	+2
Ability to analyze and solve problems	-1	0	+1	+2	-1	0	+1	+2	-1	0	+1	+2
Ability to do research	-1	0	+1	+2	-1	0	+1	+2	-1	0	+1	+2
Ability to think creatively	-1	0	+1	+2	-1	0	+1	+2	-1	0	+1	+2
Ability to continue to learn new things	-1	0	+1	+2	-1	0	+1	+2	-1	0	+1	+2
Knowledge of the requirements of your chosen profession	-1	0	+1	+2	-1	0	+1	+2	-1	0	+1	+2
Self-insight	-1	0	+1	+2	-1	0	+1	+2	-1	0	+1	+2
Attitudes and Traits												
Certainty that you have chosen the right career	-1	0	+1	+2	-1	0	+1	+2	-1	0	+1	+2
High aspirations for your career	-1	0	+1	+2	-1	0	+1	+2	-1	0	+1	+2
Personal involvement in your field	-1	0	+1	+2	-1	0	+1	+2	-1	0	+1	+2
Positive attitude toward further education	-1	0	+1	+2	-1	0	+1	+2	-1	0	+1	+2
Overall breadth of perspective, vision	-1	0	+1	+2	-1	0	+1	+2	-1	0	+1	+2
Tolerance of other people and their points of view	-1	0	+1	+2	-1	0	+1	+2	-1	0	+1	+2
Ability to work with other people	-1	0	+1	+2	-1	0	+1	+2	-1	0	+1	+2
Ability to communicate your ideas to other people effectively	-1	0	+1	+2	-1	0	+1	+2	-1	0	+1	+2
Ability to induce change in others and in organizations	-1	0	+1	+2	-1	0	+1	+2	-1	0	+1	+2
Willingness to be influenced by others	-1	0	+1	+2	-1	0	+1	+2	-1	0	+1	+2
Leadership ability	-1	0	+1	+2	-1	0	+1	+2	-1	0	+1	+2
Leadership desire	-1	0	+1	+2	-1	0	+1	+2	-1	0	+1	+2
Overall self-confidence	-1	0	+1	+2	-1	0	+1	+2	-1	0	+1	+2

12. Please rate the contribution of your pre-college experiences, your undergraduate education, your graduate or professional education, and your post-education job experiences to the development of each of the areas listed below. Use the following scale (please insert the appropriate number in each space indicated):

0	1	2
no contribution	some contribution	a significant contribution

	Pre-college experiences	Undergraduate education	Graduate education	Post-education job experiences
General attitudes and values	______	______	______	______
Professional and ethical standards . . .	______	______	______	______
Attitudes and personal characteristics needed in your profession.	______	______	______	______

13. PRESENT JOB: What is your full title in this job? ______________________

Please describe your functions in this job: ______________________

How long have you held this job? ________ How long do you think you will remain with this organization? ____________

14. Using the loose sheet marked "Lists of Job Characteristics," please describe as well as you can the following characteristics of your present job. Look over the indicated list and write the appropriate number or numbers in the space provided.

Type of organization: (List A) ______ Location: (List B) ______ Job description(s): (List C) ____________

Size of total organization: (List D) ______ Size of your department: (List D) ______ Number of people working for you: ____________

Starting salary: ($/year) ____________ Salary now: ($/year) ____________

Professional income OTHER THAN SALARY: ($/year) ____________
(e.g. consulting fees, royalties, etc.)

15. How satisfied are you with your present job? (circle the appropriate number in the scale)

1 2 3 4 5
very dissatisfied — very satisfied

16. OCCUPATIONAL HISTORY: One of the major purposes of this survey is to find out what kinds of patterns there are in people's jobs after college. We therefore would like some basic information about each job you have held since college or graduate school, and the length of time you held that job. If you have been in your present job since the beginning, check here _____, skip this section, and go to question 17.

A promotion within an organization does not constitute a new job, unless your function changed in some basic way along with the promotion. List only jobs which you held for six months or more, unless you consider some shorter job of special importance. Space is provided for five jobs (excluding your present one, which you have already described). If you have held more than five jobs before your present one, please describe the first five in the spaces provided and add a sheet which describes the others.

16.1 Initial job after school

Type of organization: (List A) ______ Location: (List B) ______ Job description(s): (List C) ____________

Size of total organization: (List D) ______ Size of your department: (List D) ______ Number of people working for you: ____________

Starting Salary: ($/year) ____________ Salary at Termination: ($/year) ____________ ____________

Professional income OTHER THAN SALARY: ($/year) ____________
(e.g. consulting fees, royalties, etc.)

Length of time job was held: ________ Reason(s) for leaving: (List E) ____________

LISTS OF JOB CHARACTERISTICS

LIST A

Type of Organization

1. Family business
2. Own professional office
3. Founder or co-founder of a business enterprise
4. Member or partner in a professional office
5. Private profit-making company or corporation
6. University research organization or affiliated institute
7. University or college: academic department or administration
8. Junior college: academic department or administration
9. Non-profit research organization or institute, NOT affiliated with a university
10. Hospital or clinic
11. Public welfare organization
12. Private welfare organization
13. Elementary or secondary school
14. Federal government
15. State government
16. Local government
17. Church or other religious organization
18. Other (please specify on questionnaire)

LIST B

Location

1. New England: (Maine, New Hampshire, Vermont, Connecticut, Massachusetts, Rhode Island)
2. Mid-Atlantic: (New York, Pennsylvania, New Jersey, Maryland, Delaware, District of Columbia)
3. South: (Virginia, North Carolina, South Carolina, Georgia, Florida, West Virginia, Tennessee, Alabama, Mississippi, Arkansas, Louisiana, Kentucky)
4. Mid-West: (Ohio, Michigan, Indiana, Illinois, Wisconsin, Minnesota)
5. North Central: (Iowa, Kansas, Nebraska, Montana, North Dakota, South Dakota, Colorado, Wyoming, Missouri, Idaho, Utah, Nevada)
6. Southwest: (Texas, New Mexico, Arizona, Oklahoma)
7. West Coast: (California, Oregon, Washington, Hawaii, Alaska)
8. Outside of the United States

LIST C

Job Descriptions

From the list of job descriptions below, pick ONE OR MORE that best describe your job. For example, if you are the director of an engineering consulting group, pick 1, 7, 11.

1. Engineer
2. Scientist
3. Architect
4. Doctor
5. Lawyer
6. Social Scientist
7. Manager, director, or executive
8. Business staff (marketing, finance, production, etc.)
9. Salesman
10. Teacher
11. Consultant
12. Research
13. Social Service
14. Other (please specify on questionnaire)

LIST D

Size

1. Less than 10
2. 10 - 99
3. 100 - 499
4. 500 - 2,499
5. 2,500 - 9,999
6. 10,000 and over

LIST E

Reasons for Changing Jobs

Please read the list of possible reasons for changing jobs given below, and then choose as many as you feel are important and LIST THEM IN ORDER OF IMPORTANCE. Each reason represents both a negative reason for leaving a job and a positive one for taking a new job. For example, you may have left a job because the workload was not reasonable, or because the workload in the new job was better. In either case, you would list reason #1.

1. Did not have a reasonable workload
2. Did not have an opportunity for advancement
3. Did not have good physical working conditions
4. Did not have good fringe benefits
5. Did not have job security (steady work)
6. Did not have an opportunity for high earnings
7. Did not like the location of the job
8. Did not have an opportunity to exercise leadership
9. Did not get the recognition I deserved when doing a good job
10. Did not work for a highly regarded company
11. Did not work in an efficiently run department
12. Did not have a job that is highly regarded by others in the company -- a job with some prestige
13. Did not have a job that allowed me to make a real contribution to the success of the company
14. Did not have challenging work to do -- work from which I can get a personal sense of accomplishment
15. Did not have enough freedom to adopt my own approach to the job -- to be creative and original
16. Did not have training or educational opportunity (to improve knowledge and skills)
17. Did not have a job that allows me to make a real contribution to society
18. Did not have a job that allows me to work on crucial, relevant problems
19. Did not work in a department where the people are friendly and congenial
20. Did not have the opportunity to work with people rather than things
21. Did not have a job that leaves sufficient time for family and personal life
22. Not by my choice (e.g. was laid off, company was shut down, job or project was finished, contract not renewed, was not re-appointed, etc.)
23. Change in family circumstances (please specify on questionnaire)
24. Other (please specify on questionnaire)

16.2 Second job after school

Type of organization: (List A) ______
Location: (List B) ______
Job description(s): (List C) ______

Size of total organization: (List D) ______
Size of your department: (List D) ______
Number of people working for you: ______

Starting Salary: ($/year) ______
Salary at Termination: ($/year) ______

Professional income OTHER THAN SALARY: ($/year) ______
(e.g. consulting fees, royalties, etc.)

Length of time job was held: ______
Reason(s) for leaving: (List E) ______

16.3 Third job after school

Type of organization: (List A) ______
Location: (List B) ______
Job description(s): (List C) ______

Size of total organization: (List D) ______
Size of your department: (List D) ______
Number of people working for you: ______

Starting Salary: ($/year) ______
Salary at Termination: ($/year) ______

Professional income OTHER THAN SALARY: ($/year) ______
(e.g. consulting fees, royalties, etc.)

Length of time job was held: ______
Reason(s) for leaving: (List E) ______

16.4 Fourth job after school

Type of organization: (List A) ______
Location: (List B) ______
Job description(s): (List C) ______

Size of total organization: (List D) ______
Size of your department: (List D) ______
Number of people working for you: ______

Starting Salary: ($/year) ______
Salary at Termination: ($/year) ______

Professional income OTHER THAN SALARY: ($/year) ______
(e.g. consulting fees, royalties, etc.)

Length of time job was held: ______
Reason(s) for leaving: (List E) ______

16.5 Fifth job after school

Type of organization: (List A) ______
Location: (List B) ______
Job description(s): (List C) ______

Size of total organization: (List D) ______
Size of your department: (List D) ______
Number of people working for you: ______

Starting Salary: ($/year) ______
Salary at Termination: ($/year) ______

Professional income OTHER THAN SALARY: ($/year) ______
(e.g. consulting fees, royalties, etc.)

Length of time job was held: ______
Reason(s) for leaving: (List E) ______

17. The list below shows a number of characteristics of a job. Please circle the appropriate number to show how important you feel each characteristic is to you with regard to your present and future jobs.

	Not at all important				Very important
Reasonable workload	1	2	3	4	5
Opportunity for advancement	1	2	3	4	5
Department where people are friendly and congenial	1	2	3	4	5
Challenging work to do	1	2	3	4	5
Work from which I could get a personal sense of accomplishment	1	2	3	4	5
Highly regarded organization	1	2	3	4	5
Recognition for doing a good job	1	2	3	4	5
Job which allows me to make a real contribution to the success of the organization	1	2	3	4	5
Good physical working conditions	1	2	3	4	5
Training or educational opportunities (to improve my knowledge or skills)	1	2	3	4	5
Efficiently run department	1	2	3	4	5
Considerable freedom to adopt my own approach to the job -- to be creative and original	1	2	3	4	5
Job regarded highly by others in the company -- a job with some prestige	1	2	3	4	5
Good fringe benefits	1	2	3	4	5
Job which leaves sufficient time for family and personal life	1	2	3	4	5
Work that is relevant to social problems	1	2	3	4	5
Job security (steady work)	1	2	3	4	5
Opportunity for high earnings	1	2	3	4	5
Location	1	2	3	4	5
Opportunity to exercise leadership	1	2	3	4	5
Job which allows me to make a contribution to society	1	2	3	4	5
Opportunity to work with people rather than with things	1	2	3	4	5
Other (please specify) ____________	1	2	3	4	5

18. What kind of a job do you expect to have at the height of your career -- at the time when you are functioning most fully in your professional work? Please describe such a job by answering the following questions, referring to the lists, where indicated, as before.

Type of organization: ____________ (List A)

Job Description(s): ____________ (List C)

Number of people working for you: ____________

Salary (today's $/year): ____________

Professional income OTHER THAN SALARY (today's $/year): ____________

	Unsuccessful				Very successful
19. At this point in your professional life, how successful do you think you are in your work?	1	2	3	4	5
How successful do you think you will be at the height of your career?	1	2	3	4	5

	Unimportant				Very important
20. How important is it to you to be successful in your work?	1	2	3	4	5

	Hardly anything				A great deal
21. Overall, how much has your UNDERGRADUATE education contributed to your success in your work?	1	2	3	4	5

22. YOUR PROFESSION: What do you consider to be your profession? ____________________

23. Which three of the following reasons for belonging to a professional society seem to you to be most important? Place a 1 next to the one you consider to be the most important reason for belonging; a 2 next to the reason you would consider to be next most important; and a 3 next to the third most important reason for belonging to a professional society.

	RANK
To get information via papers, journals, etc.	______
To meet an expectation of one's employer.	______
To help one identify with the profession.	______
To attend meetings at different locations	______
To make contacts with people professionally helpful for work.	______
To make contacts that might be helpful should one want to change jobs	______
To make social contacts	______
Other (please specify) ____________________	______

For the following sets of questions, please circle the appropriate number:

		Yes	No
24.1	Do you presently belong to a professional society?	1	2
24.2	Have you ever read a paper at a meeting of a professional society?	1	2
24.3	Have you ever published any professional articles, papers, or books?	1	2
25.1	I am more concerned with how my work looks to my professional colleagues than to my boss.	1	2
25.2	If there are conflicts between professional standards and the interests of my employer I tend to resolve them in favor of my employer	1	2

26. Please indicate the extent to which you agree with the following statements by circling the appropriate number.

	Strongly disagree				Strongly agree
If I had to change the kind of work I do, I would be very frustrated and unfulfilled.	1	2	3	4	5
I like to think about my work, even when off the job	1	2	3	4	5
My only interest in my job is to get enough money to do the other things that I want to do.	1	2	3	4	5
I wish I were in a completely different occupation	1	2	3	4	5
My main satisfactions in life come from the work I do.	1	2	3	4	5

27. People vary in the extent to which they are in agreement with the main trends of their profession. Please indicate where you place yourself on this dimension by circling the appropriate number on the following scale:

1	2	3	4	5
complete disagreement				complete agreement

28. All other things being equal, which of the following would suit you best? Circle the number of the one you would most prefer.

To work at the core of a well established field or profession 1

To work at the frontiers of a well established field or profession. 2

To work in an emergent, more nebulous, or rapidly changing field or profession. 3

29. SELF-ASSESSMENT: Below is the list of abilities and traits that you have already marked. This time, please indicate the degree to which you NOW possess each of the listed factors.

Rate yourself by circling the number from 0 to 4 that best describes the extent to which you now possess each ability or trait.

Knowledge and Abilities	do not possess at all				possess to a great extent
Real competence in your chosen field	0	1	2	3	4
Ability to identify problems	0	1	2	3	4
Ability to analyze and solve problems	0	1	2	3	4
Ability to do research	0	1	2	3	4
Ability to think creatively	0	1	2	3	4
Ability to continue to learn new things	0	1	2	3	4
Knowledge of the requirements of your chosen profession	0	1	2	3	4
Self-insight	0	1	2	3	4
Attitudes and Traits					
Certainty that you have chosen the right career	0	1	2	3	4
High aspirations for your career	0	1	2	3	4
Personal involvement in your field	0	1	2	3	4
Positive attitude toward further education	0	1	2	3	4
Overall breadth of perspective, vision	0	1	2	3	4
Tolerance of other people and their points of view	0	1	2	3	4
Ability to work with other people	0	1	2	3	4
Ability to communicate your ideas to other people effectively	0	1	2	3	4
Ability to induce change in others and in organizations	0	1	2	3	4
Willingness to be influenced by others	0	1	2	3	4
Leadership ability	0	1	2	3	4
Leadership desire	0	1	2	3	4
Overall self-confidence	0	1	2	3	4

30. At what point in your career were you married and was your first child born?

	Married	First child born
While an undergraduate in college	1	1
Between college and graduate school	2	2
While in graduate or professional school	3	3
During the first two years at work	4	4
After working two years or more	5	5
Other (please specify) ______	6	6
Not applicable	7	7

31. Which three aspects of your life give you the most satisfaction? In the following list, place a 1 next to the item that gives you the most satisfaction in life; a 2 next to the one that gives you the next most satisfaction; and a 3 next to the third most satisfying aspect of your life.

	RANK
Career or occupation	________
Creative or other activities not related to career or occupation	________
Leisure time recreational activities	________
Family relationships	________
Activities directed at community, national, or international betterment. . .	________

32. Here are six descriptions of the relationship between work and family. Please indicate (by circling):

A. Which is closest to yours now?

B. Which is the one you would ideally like to have now?

C. Which is the one you would like to have at the height of your career?

	Now	Ideally now	At height
Single person living alone (or with friends) and working	1	1	1
Married, husband works and wife is housewife	2	2	2
Married, wife works while husband runs the home.	3	3	3
Married, husband and wife work at different kinds of jobs.	4	4	4
Married, husband and wife work at related kinds of work but not together (e.g., two architects in different practices)	5	5	5
Married, husband and wife work together.	6	6	6
Another arrangement (please specify) ________________________	7	7	7

33. For each of the following periods in your family life, please indicate whether YOUR WIFE worked or plans to work full-time, part-time, or not at all. (If you are not married, please answer according to what you think your wife would choose to do if you were married.)

	Full time	Part time	Not at all	Don't know
After marriage but before birth of first child	1	2	3	4
When youngest child is under 3 years old	1	2	3	4
When youngest child is 6 - 12 years old.	1	2	3	4
After all children are grown	1	2	3	4

34. In retrospect, what has been most disappointing to you about your undergraduate education at M.I.T.?

35. Knowing what you do now, what changes, if any, in your undergraduate or graduate education would have been beneficial for you?

36. What kind of continuing educational opportunities do you think M.I.T. should make available to its alumni?

Bibliography

Athanasiou, R. 1971. "Selection and socialization: A study of engineering student attrition." *Journal of Educational Psychology* 62: 157–166.

Bailyn, L. 1970. "Career and family orientations of husbands and wives in relation to marital happiness." *Human Relations* 23: 97–113.

Bailyn, L. 1973. "Family constraints on women's work." *Annals of the New York Academy of Science* 208: 82–90.

Bailyn, L. 1977a. "Involvement and accommodation in technical careers: An inquiry into the relation to work at mid-career." In J. Van Maanen, ed. *Organizational Careers: Some New Perspectives.* New York: Wiley International.

Bailyn, L. 1977b. Technical careers at mid-stream: An analysis of mid-career issues and their organizational implications. MIT Sloan School of Management Working Paper 930-77.

Bailyn, L. 1977c. "Research as a cognitive process: Implications for data analysis." *Quality and Quantity* 11: 97–117.

Bailyn, L. 1978. "Accommodation of work to family." In R. Rapoport and R. N. Rapoport, eds. *Working Couples.* London: Routledge and Kegan Paul.

Bailyn, L. 1979. "Taking off for the top: How much acceleration for career success?" *Management Review* 68 (1): 18–23.

Bailyn, L., and Schein, E. H. 1972. "Where are they now and how are they doing?" *Technology Review* 74 (7): 3–11.

Bennis, W. 1976. *The Unconscious Conspiracy: Why Leaders Can't Lead.* New York: AMACOM.

Blauner, R. 1964. *Alienation and Freedom: The Factory Worker and His Industry.* Chicago: University of Chicago Press.

Cohen, A. R., and Gadon, H. 1978. *Alternative Work Schedules: Integrating Individual and Organizational Needs.* Reading, Mass.. Addison-Wesley.

Dalton, D. W., and Thompson, P. H. 1971. "Accelerating obsolescence of older engineers." *Harvard Business Review* 49 (September–October): 57–67.

David, D. S. 1971. *Career Patterns and Values: A Study of Men and Women in Science and Engineering.* New York: Bureau of Applied Social Research, Columbia University.

Davis, J. A. 1965. *Undergraduate Career Decisions: Correlates of Occupational Choice.* Chicago: Aldine.

Dipboye, W. J., and Anderson, W. F. 1961. "Occupational stereotypes and manifest needs." *Journal of Counseling Psychology* 8: 296–304.

Erikson, E. H. 1959. "Identity and the life-cycle." *Psychological Issues* 1 (1): 1–177.

Ferdinand, T. N. 1966. "On the obsolescence of scientists and engineers." *American Scientist* 54 (1): 46–56.

Fogarty, M. P., Rapoport, R., and Rapoport, R. N. 1971. *Sex, Career and Family.* London: Allen & Unwin.

Glaser, B. G. 1964. *Organizational Scientists: Their Professional Careers.* New York: Bobbs-Merrill.

Gordon, J. R. 1977. The congruence between the job orientation and job content of management school alumni. Ph.D. diss., MIT Sloan School of Management.

Gorsuch, R. L. 1974. *Factor Analysis.* Philadelphia: Saunders.

Gould, R. L. 1978. *Transformations: Growth and Change in Adult Life.* New York: Simon and Schuster.

Gouldner, A. W. 1957. "Cosmopolitans and locals: Toward an analysis of latent social roles–I." *Administrative Science Quarterly* 2: 281–306.

Gouldner, A. W. 1958. "Cosmopolitans and locals: Toward an analysis of latent social roles–II." *Administrative Science Quarterly* 2: 444–480.

Graves, J. P. 1978. Effective and successful management: Contrary effects on subordinate career development. Paper presented at 1978 MIT Career Research Conference. To be published in R. Katz, ed. *Career Issues for Human Resource Management.* Englewood Cliffs, N.J.: Prentice-Hall.

Haberman, S. J. 1973. "The analysis of residuals in cross-classified tables." *Biometrics* 29: 205–220.

Herzberg, F. 1966. *Work and the Nature of Man.* Cleveland: World.

Holland, J. L. 1959. "A theory of vocational choice." *Journal of Counseling Psychology* 6: 35–45.

Holland, J. L. 1966. *The Psychology of Vocational Choice: A Theory of Personality Types and Environmental Models.* London: Ginn.

Holmstrom, L. L. 1972. *The Two-Career Family.* Cambridge, Mass.: Schenkman.

Hudson, L. 1967. *Contrary Imaginations: A Psychological Study of the English Schoolboy.* Harmondsworth, England: Penguin.

Izard, C. E. 1960. "Personality characteristics of engineers as measured by the EPPS." *Journal of Applied Psychology* 44: 332–335.

Jacobson, R. C. 1977. The socialization of technology trained college hires in a computer company. Master's thesis, MIT Sloan School of Management.

Johnson, E. C., Jr. 1975. Engineers: Interdisciplinary nonhomogeneity and job satisfaction. Master's thesis, MIT Sloan School of Management.

Kahn, R. L. 1972. "The meaning of work: Interpretation and proposals for measurement." In A. Campbell and P. Converse, eds. *The Human Meaning of Social Change*. New York: Russell Sage Foundation.

Kanter, R. M. 1977. "Some effects of proportions on group life: Skewed sex ratios and responses to token women." *American Journal of Sociology* 82: 965–990.

Kolb, D., and Fry, R. 1974. Towards an applied theory of experiential learning. MIT Sloan School of Management Working Paper 732-74.

Kolb, D., and Goldman, M. 1973. Toward a typology of learning styles and learning environments: An investigation of the impact of learning styles and discipline demands on the academic performance, social adaptation and career choices of MIT seniors. MIT Sloan School of Management Working Paper 688-73.

Kubie, L. 1953. "Some unsolved problems of the scientific career." *American Scientist* 41: 596–613.

Kubie, L. 1954. "Psychoneurotic problems of the American scientist." *Chicago Review* 8: 65–80.

Kuder, G. F. 1957. "A comparative study of some methods of developing occupational keys." *Educational and Psychological Measurement* 17: 105–114.

Laws, J. L. 1975. "The psychology of tokenism: An analysis." *Sex Roles* 1: 51–67.

Levinson, D. J., et al. 1978. *The Seasons of a Man's Life*. New York: Knopf.

Madrazo, A. 1974. Personal and environmental factors affecting engineer career development. Master's thesis, MIT Sloan School of Management.

Marquis, D. 1965. Scientists in a technology-oriented organization—Their expectations, incentives, & career patterns. Paper presented at Institute on Research Administration, Washington, D.C.

McKelvey, B., and Sekaran, U. 1977. "Towards a career-based theory of job involvement: A study of scientists and engineers." *Administrative Science Quarterly* 22: 281–305.

Merton, R. K. 1949. "Patterns of influence: A study of interpersonal influence and of communications behavior in a local community." In P. F. Lazarsfeld and F. Stanton, eds. *Communications Research 1948–49.* New York: Harper.

Moore, H., and Levy, S. H. 1951. "Artful contriver: A study of engineers." *Personnel* 28: 148–153.

Morse, N. C., and Weiss, R. S. 1955. "The function and meaning of work and the job." *American Sociological Review* 20: 191–198.

Nie, N. H., Hall, C. H., Jenkins, J. G., Steinbrenner, K., and Bent, D. H. 1975. *Statistical Package for the Social Sciences,* second edition. New York: McGraw-Hill.

Osipow, S. H. 1973. *Theories of Career Development.* New York: Appleton-Century-Crofts.

Pearse, R. F., and Pelzer, B. P. 1975. *Self-Directed Change for the Mid-Career Manager.* New York: AMACOM.

Pelz, D. C., and Andrews, E. M. 1966. *Scientists in Organizations: Productive Climates for Research and Development.* New York: Wiley.

Perrucci, R., and Gerstl, J. E. 1969. *Profession Without Community: Engineers in American Society.* New York: Random House.

Pigors, P., and Myers, C. A. 1977. *Personnel Administration,* eighth edition. New York: McGraw-Hill.

Plovnick, M. S. 1972. Social awareness and role innovation in engineers. MIT Sloan School of Management Working Paper 603-72.

Rabinowitz, S. and Hall, D. T. 1977. "Organizational research on job involvement." *Psychological Bulletin* 84: 265–288.

Rapoport, R., and Rapoport, R. N. 1976. *Dual-Career Families Re-examined: New Integrations of Work and Family.* New York: Harper and Row.

Rapoport, R. N. 1970. *Mid-Career Development: Research Perspectives on a Developmental Community for Senior Administrators.* London: Tavistock.

Rapoport, R. N., and Lohman, E. O. No date. Technologists in mid-career: Factors affecting the "ten-year-out" patterns of engineers and scientists from three universities. Unpublished draft.

Reiss, H., and Balderston, J. 1966. "Motivating scientists." *International Science and Technology* 5 (June): 93–99.

Reynolds, H. T. 1977. *The Analysis of Cross-Classifications.* New York: Free Press.

Rhoades, R. G., Roberts, E. B., and Fusfeld, A. R. 1978. "A correlation of R&D laboratory performance with critical functions analysis." *R&D Management* 9: 13–17.

Riesman, D. 1970. "Observations on contemporary college students, especially women." *Interchange* 1: 52–63.

Ritti, R. 1971. *The Engineer in the Industrial Corporation.* New York: Columbia University Press.

Roberts, E. B, and Wainer, M. A. 1971. "Some characteristics of technical entrepreneurs." *IEEE Transactions in Engineering Management* 18: 100–109.

Roe, A. 1956. *The Psychology of Occupations.* New York: Wiley.

Roe, A. 1957. "Early determinants of vocational choice." *Journal of Counseling Psychology* 4: 212–217.

Roe, A. 1961. "The psychology of the scientist." *Science* 134: 456–59.

Rosenberg, M. 1957. *Occupations and Values.* Glencoe, Ill.: Free Press.

Schein, E. H. 1972. "The general manager: A profile." In *Proceedings of the Eastern Academy of Management.*

Schein, E. H. 1978. *Career Dynamics: Matching Individual and Organizational Needs.* Reading, Mass.: Addison-Wesley.

Sheehy, G. 1976. *Passages: Predictable Crises of Adult Life.* New York: Dutton.

Shepherd, C. R. 1961. "Orientations of scientists and engineers." *Pacific Sociological Review* 4: 79–83.

Shuttleworth, F. K. 1940. Sampling errors involved in incomplete returns to mail questionnaires. Paper presented at the 1940 annual meeting of the American Psychological Association, Pennsylvania State College. Summarized in *Psychological Bulletin* 37 (1940): 437.

Sirota, D., and Greenwood, J. M. 1971. "Understand your overseas workforce." *Harvard Business Review* 49 (January–February): 53–60.

Sofer, C. 1970. *Men in Mid-Career: A Study of British Managers and Technical Specialists.* Cambridge University Press.

Steiner, M. 1953. "The search for occupational personalities: The Rorschach test in industry." *Personnel* 29: 335–343.

Sternberg, C. 1955. "Personality trait patterns of college students majoring in different fields." *Psychological Monographs* 69 (whole no. 403): 1–21.

Strong, E. K., Jr. 1943. *Vocational Interests of Men and Women.* Stanford, Calif.: Stanford University Press.

Super, D. E., et al. 1963. *Career Development: A Self-Concept Theory,* Research Monograph 4. New York: College Entrance Examination Board.

Thompson, P. H., and Dalton, G. W. 1976. "Are R&D organizations obsolete?" *Harvard Business Review* 54 (November–December): 105–116.

Underhill, R. 1967. *Occupational Values and Post-College Choice.* Chicago: National Opinion Research Center.

Vaillant, G. E. 1977. *Adaptation to Life.* Boston: Little, Brown.

Vroom, V. H. 1962. "Ego-involvement, job satisfaction and job performance." *Personnel Psychology* 15: 159–177.

Wilensky, H. L. 1961. "Orderly careers and social participation: The impact of work history on social integration in the middle mass." *American Sociological Review* 26: 521–539.

Wilensky, H. L. 1964. "Varieties of work experience." In H. Borow, ed. *Man in a World at Work.* Boston: Houghton-Mifflin.

Yankelovich, D. 1972. *The Changing Values on Campus.* New York: Pocket Books.

Index

Accommodative careers, 59, 63, 64, 65, 77
of women graduates, 58, 59, 60, 61, 63–64
Adult development, 1, 4
Alienation, xi, 1, 33, 106n.5. *See also* Work involvement
Architects, 12, 14, 15
Athanasiou, R., study of engineering students by, 28
Autonomy, 22, 28, 47, 54

Bennis, W.
on "scanner," 72, 76–77
on "variance sensor," 72, 76–77
Business staff, 19, 42, 94
definition of, 12
and work involvement, 36, 41

Career, 98, 106n.2. *See also* Early career; Mid-career; Orderly career; Technical careers
development of, 1–2
meaning of, 22, 23 (*see also* Work, meaning of)
slow periods in, 63, 66n.3, 79
Career accommodations, 8, 15, 57–58, 79. *See also* Accommodative careers
Career orientations. *See* Orientations to work
Career outcomes. *See* Job satisfaction; Perceived success; Work involvement
Career paths, 67, 79. *See also* Engineering-based career pattern; Engineering career path; Scientific-professional career pattern
technical vs. managerial, 4–5, 47, 72–73 (*see also* Technical careers)
of women graduates, 58–59, 61, 63
Career patterns, 7, 30–31. *See also* Engineering-based career pattern; Scientific-professional career pattern
defined by objective career data, 19–22, 68
in relation to value norms, 7, 19, 22, 31, 39, 68–69 (*see also* Job values)
Career styles of women graduates, 58–59, 61. *See also* Women graduates, degree of career participation
accommodative careers, 63–64
full careers, 61–63
married vs. single, 62–63
no careers, 64
Career success. *See* Perceived success; Success
Career types, 8, 38, 69. *See also* Engineering-based career pattern; Scientific-professional career pattern
characteristics of, 42, 43–44, 53–55
defined by career pattern and organizational position, 8, 33–34, 38, 69
fit with individual orientations, 8, 47, 50–53, 70
illustrated, 38–39
in relation to job values and success aspirations, 39–40, 45–46n.3, 46n.4
in relation to orientations to work, 49–50
in relation to work involvement, 39–41
Career values. *See* Job values
Challenge of work, 1, 47. *See also* Work, meaning of
Children, and careers, 65. *See also* Women graduates
Computer applications, 12, 15
Consultants, 12, 15, 19, 42, 53, 94
and work involvement, 41
Cosmopolitan vs. local orientation, 107, 108

Daedalus, on limits of scientific inquiry, 9n.1

Dalton, G. W. *See* Thompson, P. H.
David, D. S., on "value indexes," 31n.2
Davis, J. A., on dimensions of values, 31n.2
Dual-career families, 34–35, 64, 65–66
Dual ladder, 4, 72

Early career, 30–31, 65, 66n.3, 79. *See also* Women graduates, early career experiences of
engineering-based, 19
scientific-professional, 20
Education, employment in, 12, 14
Employee effectiveness. *See* Performance
Employment sector, 17, 84–85
Engineering, 17
as lifetime occupation, 69–70, 73
Engineering-based career patterns, 7, 20, 22, 27–28, 31, 39, 43–44, 70
defined, 19
with high organizational position, 42, 52, 53–54, 69, 70
with low organizational positions, 42, 44, 48, 52, 54, 69, 70
Engineering career path, 19, 21–22, 53
Engineering managers, 16, 17, 19, 20, 42, 94
and work involvement, 41
Engineering professors, 17, 21–22, 26, 68, 94
and work involvement, 36, 41
Engineers, staff, 9n.2, 16, 17, 19, 20, 25, 42, 44, 95
definition of, 12
and job values, 46n.6
negative reactions to work, 54
as different from scientists, 15, 21, 31n.1, 69
and work involvement, 36, 41
if technically oriented, 41
Entrepreneurs, 16, 19, 20, 42, 53, 94, 95, 96, 98
definition of, 11
and work involvement, 40
Erikson, E., on "generativity," 4
Expectations, about life and career, 1, 3, 33
Extrinsic job characteristics, 23, 103, 105. *See also* Job values

Family, orientation toward, 3, 48, 76, 98. *See also* Orientations to work, non-work
in measure of work involvement, 34–35
of women graduates, 62–64,
Family status, 17, 84–85
of women graduates, 59–60
Father's occupation, 16–17, 18, 84–85
Ferdinand, T. N., on "areal" obsolescence, 74–75
First-line supervisors, 12
Fit, of individuals to jobs, 44, 47, 48, 53, 54, 55, 75. *See also* Career types, fit with individual orientations
Fry, R. *See* Kolb, D.
Functional managers, 16, 19, 42, 53, 94, 95
definition of, 11
and work involvement, 41

General managers, 9n.2, 16, 18, 19, 42, 53, 56n.4, 68, 94
definition of, 12
and work involvement, 40
Gouldner, A. W., on cosmopolitan vs. local orientation, 107
Graves, J. P., on effective vs. successful managers, 72, 76

Hall, D. T. *See* Rabinowitz, S.
Herzberg, F., on two-factor theory of job satisfaction, 106n.5
Human orientation. *See* Orientations to work, to people
of women graduates, 62, 63

Income, 47, 52, 54, 55
differences among occupational roles, 17, 18, 84–85
of employed women graduates, 61, 62, 63
normalized, as indicator of performance, 50, 55–56n.3, 71
Index of work involvement, 34, 35–36, 44–45n.1, 100. *See also* Work involvement
derivation, 97–101
item analysis, 100–101
as measure of non-work orientation, 109
in relation to intrinsic/extrinsic job characteristics, 103–105
Intrinsic job characteristics, 23, 26–28, 33, 100, 103. *See also* Job values

Job classification. *See* Occupational classification for technology-based jobs
Job satisfaction, 2, 7, 100, 106n.2
and career types, 42–43, 51–53
of women with accommodative careers, 64
Job values, 19, 31n.2, 68–69. *See also* Career patterns, in relation to value norms; Career types, in relation to job values and success aspirations
development of, 28–30
differences between scientists and engineers, 22–23
engineering syndrome, 25–26, 31, 68
question used to define, 23, 31n.3
scientific-professional syndrome, 26–27, 31, 68–69
of women with accommodative careers, 63–64

Kahn, R. L., 45n.1
Kolb, D., and Fry, R., study on cognitive style, 28

Lohman, E. O. *See* Rapoport, R. N.

McKelvey, B., and Sekaran, U., on job involvement of scientists and engineers, 41
Management
definition of, 11
of human resources, 79
of people vs. technical problems, 48
of technical personnel, 5–6, 71, 72–73, 78
Managerial jobs, 11. *See also* Occupational classification for technology-based jobs
Managers, 14, 17–18. *See also* Engineering managers; Functional managers; General managers; Science managers; Technical managers
and meaning of work, 108
response to work and family of, 34–35
"Master," 72
pairing with apprentice, 74
Matrix of ability and orientations, 71, 72
relation to psychological contract, 71
Mentor, 72, 76
Merton, R. K., on cosmopolitan vs. local orientation, 107
Mid-career, 2, 7, 31, 33, 61, 62–63, 67
in relation to early career, 65, 66n.3
variety of interests and capacities, 5, 47, 48, 71, 78
vs. younger employees, 74–75
Morse, N. S., and Weiss, R. S., 45n.1

Non-work orientation. *See* Orientations to work, non-work
Norms. *See* Job values

Obsolescence, 74–75
Occupational classification for technology-based jobs, 10–11, 67–68. *See also names of occupational roles*
 in eleven occupational ròles, 15–16 (*see also* Occupational roles)
 in two career patterns, 19–22, 68 (*see also* Career patterns)
 in four career types, 38, 69 (*see also* Career types)
Occupational norms. *See also* Job values
 determination from individual data, 23, 25
Occupational roles, 2, 7, 12–15. *See also names of occupational roles*
 differences in employment characteristics, 17, 84–85
 differences in family characteristics, 17–18, 84–85
 differences in socioeconomic origins, 16–17, 84–85
 differences by year of graduation, 14
 of employed women graduates, 62
 and success, 94–96
Occupational socialization, 28–30, 32n.10
Occupational success. *See* Perceived success; Success
Occupational values. *See* Job values
Orderly career, 63, 65–66
Organizational evaluation of effectiveness. *See* Performance
Organizational position, 8, 32, 45n.2, 47. *See also* Career types
 definition of, in relation to work involvement, 37
 relation of, to success aspirations, 37–38, 39, 69
Organizational response, creative, 5–6, 52–53, 70–71, 73–79
Organizational rewards, 5, 47, 50–51, 55, 73, 75, 76
Organizational roles, need for multiple, 5–6, 9, 55, 71–78
Organizational value syndrome, 7, 28. *See also* Job values, engineering syndrome
Orientations to work, 8, 48–50, 53–55, 55n.1, 69, 71
 non-work, 77 (*see also* Accommodative careers; Family, orientation toward)
  defined, 48, 109
  in high performers, 76–77
  in low performers, 77–78
 to people ("human orientation")
  defined, 48, 108–109
  in high performers, 75
  in low performers, 75–76
 technical
  defined, 48, 107–108
  in high performers, 71–73
  in low performers, 74–75
 of women graduates, 62, 63

Part-time work. *See* Accommodative careers
Patterning of technical occupations, 69. *See also* Technical jobs
People, orientation to. *See* Orientations to work, to people
Perceived success, 2, 46n.5. *See also* Success
 and career types, 42–43
 measure of, 93–94
 in relation to occupational role, 94–95
 of women graduates, 62, 63
Performance, 50–51, 71, 78
Personnel policies and procedures. *See* Organizational response; Organizational rewards; Organizational roles
Plateaued employees, 75
Pluralism, xii, 4–6, 79
Professionalism
 of architects, 12, 15

as differentiating career patterns, 68
of women graduates, 62, 63
Professional value syndrome, 7, 28. *See also* Job values, scientific-professional syndrome
Professors, 14, 17, 18, 19, 21, 42, 54. *See also* Engineering professors; Science professors
Psychological contract, 71

Questionnaire, 6, 34
response rate, 6

Rabinowitz, S., and Hall, D. T., 45n.2
Rapoport, R. N. and Lohman, E. O.
commitment-alienation scale of, 97, 98, 99, 104, 108
professionalism scale of, 107
Reactions to work, 5–6, 7, 8, 33. *See also* Work involvement; Job satisfaction; Perceived success
Respondents. *See* Sample
Response bias, 6, 12, 14, 89
Rhoades, R. G., on "coaching," 72, 76
Riesman, D., report on Smith survey on coeducation, 66n.2
Rosenberg, M.
on effect of values on occupational choice, 28
on three value orientations, 31n.2

Salary. *See* Income
Sample, xi, 6–7. *See also* Response bias
basic characteristics of, 10, 14–15
response rate, 6
Satisfaction. *See* Job satisfaction
Schein, E. H., panel study of M.B.A.s by, 28
Science, 2, 17
Science managers, 17, 19, 25–26, 27, 42, 54, 94, 95–96
characterization of, 20–21
and work involvement, 36, 40
Science professors, 16–17, 26, 94, 95, 96
Scientific-professional career pattern, 7, 20, 22, 27–28, 31, 39, 44, 69, 70
definition of, 19–20
with high organizational positions, 42, 54–55, 69
with low organizational positions, 42, 55, 55n.2
Scientists, staff, 17, 18, 20, 25, 42, 55, 94, 98
definition of, 12
difference from engineers, 15, 21, 31n.1, 69
and work involvement, 41
Sekaran, U. *See* McKelvey, B.
Social change, 1, 14
in attitudes and values, 65, 77, 79
Social values, 2, 14, 32n.5, 33
Sofer, C., on measure of work involvement, 45n.2
Staff engineer. *See* Engineers, staff
Staff jobs, 14
definition of, 11, 12
Staff scientist. *See* Scientists, staff
Success, 2, 3, 7, 65. *See also* Perceived success
objective and subjective measures of, 93, 95–96
Success aspirations, 8, 37–38, 39, 42–43, 69. *See also* Career types, in relation to job values and success aspirations
"Successful" respondents, 32n.6. *See also* Perceived success

Technical careers, 1, 5, 9, 52–53, 54, 67–68, 79. *See also* Engineering-based career pattern; Scientific-professional career pattern

Technical education, 5
Technical jobs, 14
 structure of, in organizations, xi, 54, 67, 74
Technical managers, 53. *See also* Engineering managers; Science managers
 definition of, 12
Technical orientation. *See* Orientations to work, technical
 of engineers, 41, 51–52, 53–54
 of women graduates, 62, 63
Technical tasks, routine, 52
Technologists, 12, 108
Technology as basis for careers, 2, 3, 10, 14–16, 67–68, 78. *See also* Technical careers
Thompson, P. H., and Dalton, G. W.
 on "idea innovators," 72
 on "internal entrepreneurs," 72

Undergraduate majors, 6, 89–91
 of women graduates, 58
Underhill, R., on effect of career on occupational values, 28
Unemployed respondents, 12, 14

Value syndromes. *See* Job values
Values. *See* Job values; Social values
Vroom, V. H., on "ego involvement," 45n.1

Weiss, R. S. *See* Morse, N. S.
Wilensky, H. L., on involvement vs. alienation, 106n.5
Wives
 of managers, 34
 professional and working status of, 17–18, 84–85
Women
 at MIT, 57, 66n.1
 single vs. married, 59, 62–63, 64, 65
 usefulness of MIT education to, 64
Women graduates, 8, 57–66, 67, 78–79. *See also* Career styles of women graduates
 degree of career participation, 58, 59
  and education, 60
  and family status, 59–60
  and first jobs, 60–61, 65
  and number of children, 59–60, 65
  and timing of children, 60
 early-career experiences of, 58, 60–61, 65, 66n.3
Work, meaning of, 1, 3, 35, 98, 100, 103, 106n.7. *See also* Orientations to work
Work involvement, 2, 7–8, 34, 37, 39–40, 69. *See also* Index of work involvement; *names of occupational roles*
 high vs. low groups, 35, 101–103
 not necessary for all employees, 77–78, 79

Younger employees, organizational bias toward, 4, 74–75